Classically Semisimple Rings

Martin Mathieu

Classically Semisimple Rings

A Perspective Through Modules and Categories

Martin Mathieu
School of Mathematics and Physics
Queen's University Belfast
Belfast, UK

ISBN 978-3-031-14208-6   ISBN 978-3-031-14209-3   (eBook)
https://doi.org/10.1007/978-3-031-14209-3

Mathematics Subject Classification: 16-01, 16B50, 16D10, 16D60, 16D70, 16D90

This Springer imprint is published by the registered company Springer Nature Switzerland AG
The registered company address is: Gewerbestrasse 11, 6330 Cham, Switzerland

To my children,
who never cease
to amaze me.

Le savant n'étudie pas la nature
parce que cela est utile;
il l'étudie
parce qu'il y prend plaisir
et il y prend plaisir
parce qu'elle est belle.

Henri Poincaré, 1908

Preface

This book is written for advanced undergraduate and beginning graduate students. It tells the story of a classical classification problem, about rings that are semisimple in the sense of Wedderburn and Artin. One of the first major achievements of modern algebra, the Wedderburn–Artin theorem, paved the way for many like theories in the first half of the twentieth century. We present this theory from a modern viewpoint choosing an approach via modules. At the same time, the reader is allowed a glimpse into category theory, which partly finds its origins in the theory of modules. Category theory is highly abstract and therefore sometimes hard to digest in a first helping. By interweaving the basic concepts with more concrete examples coming from modules and rings, we aim to make the abstract ideas more accessible and, at the same time, to present a classical beauty in an attractive modern setting.

Belfast, UK   Martin Mathieu
Spring 2022

Contents

Introduction

In order to understand a mathematical object—in our case, a ring—one introduces a variety of properties the object may have and distinguishes different objects according to these properties. In pursuing this, it is often helpful to let the object 'act' on another kind of object which may have a simpler structure—in our case, the ring acts on an abelian group. This births the concept of a module. A slightly different, but related, point of view is to 'represent' the object (the ring) as a subobject of a prototypical object (here, as a subring of an endomorphism ring). Thus, by placing the object of interest into a variety of settings one obtains a better view of it than by just 'staring at it' from one perspective.

One may want to think of all possible representations (modules) of our given object (ring), put them into a 'big set' and study this set. The process starts again: one introduces a variety of properties the modules may have ..., and one ends up with the need to bring some order into the situation. What can we do with the 'set of all modules' of a given ring and how can we relate different subsets to each other, or even relate those to sets made up of different structures? In this way, the notion of a functor between categories is created.

This book comprises material which formed part of lecture courses by the author given at the Eberhard Karls Universität Tübingen (a long time ago) and at Queen's University Belfast (far more recently). The aim is to introduce undergraduate students with some background in abstract algebra or beginning postgraduate students to the idea of investigating certain rings by exploring their categories of modules. We have three goals in mind: firstly to study, in all detail, a very specific class of (non-commutative) rings which became prominent in the early part of the twentieth century. Secondly, to employ the theory of modules in order to achieve this goal. Upon first impression, modules look like a slight generalisation of vector spaces, but one quickly realises that they are, in fact, a vast generalisation. For instance, the notion of a basis and the invariance of the length of a basis, available and fundamental for vector spaces (even in infinite dimensions, cf., e.g., [22, p. 241]), lose its significance. In the second half of the twentieth century, a more abstract approach to mathematics became predominant, and it emerged that the best way to handle the multitude of different types of modules is to put them into various classes and to study the relation between these classes. The theory of categories and functors which grew out of this endeavour has found its way into every niche of mathematics, as a universal language in which the essential qualities that make things work can be exhibited. The downside of such an abstract approach, however, is that one can easily be overwhelmed by terminology. Categories of modules have the advantage that we can keep our feet on the ground whilst remaining very comprehensive (in a sense that can be made precise, see Sect. 5.4 on abelian categories). Finally, therefore, the third goal of this book is to guide the reader into and through this abstract world in small, manageable steps.

The prerequisites to use this book successfully are minimal. The reader should have studied Linear Algebra (the first few chapters of [20] or [33], for example, will suffice; see also the treatment in [9]) as well as some basics in Ring Theory (e.g., [7], [23] or [34]); even the commutative theory may be sufficient. A good test is to look at the list of examples of rings on page 2. If the reader can understand (most of) them, they are well prepared for reading the book. A modest familiarity with groups (such as contained in [34, Chapter 4], for example) without any deeper theory is also helpful.

We begin with a motivation for the concept of a module in a leisurely manner in Chap. 1, by observing how modules appear in a natural way in various instances. We then have a first glimpse into the world of categories and look at a substantial list of examples in the hope of convincing the reader that they are ubiquitous in mathematics.

Chapter 2 is devoted to the basic constructions that one wants to, and has to, perform with modules in order to build up a viable theory. These include quotients, direct sums and direct products. We also have a look at the special class of free modules; these are the ones that resemble vector spaces the most. This discussion naturally leads us to consider objects with similar attributes in categories; often these are defined via universal properties. The rather short Chap. 3 discusses the fundamental isomorphism theorems for modules. Functors and mappings between functors ('natural transformations') are introduced in the second part of this chapter.

Finiteness conditions on submodules are the subject of the following two chapters. In Chap. 4, we discuss Noetherian modules and rings including Hilbert's basis theorem. The various permanence properties of Noetherian modules are dealt with using exact sequences; this also allows for a smooth transition into the second half of the chapter where a setting for exact sequences in more general categories is introduced and explored, together with the concept of exact functors. In this chapter the reader will start to realise that the morphisms are the really decisive parts of a category; this is another advantage of using the general setting of categories which emphasises this feature in contrast to the importance that usually is given to the objects in a theory, whether it is simple groups or Banach spaces, for example.

Close relatives of the Noetherian modules are the Artinian ones, which are investigated in Chap. 5. Much of their theory runs in parallel, but there are differences. For example, the notion of a finitely cogenerated module cannot be expressed in terms of elements, in contrast to the concept of finite generation of a module. Another is that commutative Artinian rings, studied in Sect. 5.2, allow for a more detailed structure theory. The highlight of this chapter is the Hopkins–Levitzki theorem (Theorem 5.3.2) stating that every left Artinian ring is left Noetherian. The converse statement is in general false: the ring $\mathbb{Z}$ is the most obvious example. The proof of this fundamental result is deferred to Sect. 7.3, but Exercise 5.5.8 already invites a proof in the commutative case. In Sect. 5.4 abelian categories are introduced which are the most prominent generalisation of module categories, and various typical properties and techniques are displayed.

The heart of the theory is Chap. 6, where simple and semisimple modules are defined and studied. Finitely generated semisimple modules turn out to be both Noetherian and Artinian (Theorem 6.2.4). The concept of a classically semisimple ring is introduced, and these rings are characterised by the fact that their category of (left) modules entirely consists of semisimple modules (Theorem 6.1.12). This, in consequence, leads to other important classes of modules, projective and injective modules (Sect. 6.2) and their generalisations to projective and injective objects in arbitrary categories (Sect. 6.3). The description via the Hom-functors opens up the connection to Homological Algebra which we do not pursue further here; see [21] or [29] for a full development.

The Artin–Wedderburn theorem is the focus of Chap. 7. This was one of the major achievements of the early theory of non-commutative rings and still is a cornerstone of basic algebra. It forms the model for many structure theorems that were obtained during the twentieth century. An important example of a semisimple ring is the group ring of certain finite groups; this is Maschke's theorem (Theorem 7.2.1) discussed in Sect. 7.2. A full proof of the Hopkins–Levitzki theorem, Theorem 5.3.2, is finally given in Sect. 7.3. This section also contains a discussion of the Jacobson radical for general unital rings and the notion of a semiprimitive ring. Our proof differs from the one commonly given, which relies on composition series, and uses module techniques instead. We do not have anything to say about categories in this chapter aside from Remark 7.1.13.

A construction of paramount importance is the tensor product of modules to which Chap. 8 is devoted. The fundamental Hom-tensor adjointness relation is established in Theorem 8.1.9 and further studied in Sect. 8.3 on adjoint functors between general categories. The special situation of tensor products of algebras over a field is the theme of Sect. 8.2.

An interesting generalisation of semisimple rings was introduced in 1972 by Warfield, based on previous work on modules by Crawley and Jónsson. In order to provide the reader with some more recent development in Ring Theory, we have included Chap. 9 which gives an introduction to the theory of exchange modules and exchange rings. More comprehensive treatments can be found in [13] and [16]. In Sect. 9.1, we compile a number of basic properties of exchange modules, and in Sect. 9.2, exchange rings are introduced via a separation property using idempotents (Definition 9.2.1). It is shown that a unital ring R is an exchange ring if and only if the standard module $_R R$ is an exchange module (Theorem 9.2.6) which in turn is equivalent to the fact that *every* projective module in the category $R-mod$ has the exchange property (Corollary 9.2.7). Commutative exchange rings are investigated in great detail in Sect. 9.3 and the reader will discover a number of equivalent properties characterising exchange rings (Theorem 9.3.10). We also make contact with various notions of dimension of a commutative ring and with topology; e.g., it is shown in Proposition 9.3.2 that a compact Hausdorff space X is zero-dimensional if and only if the ring $C_{\mathbb{R}}(X)$ of all continuous real-valued functions on X is an exchange ring.

The problem when a group ring is semiprimitive has challenged algebraists for a long time. It will be studied in Chap. 10, where we highlight some of the major techniques and results. It is interesting to see how analytic tools, explored in Sect. 10.2, lead to Rickart's result from 1950 stating that the group ring with complex coefficients is semiprimitive independent of the properties of the group (Theorem 10.3.1). We do not aim for a comprehensive discussion of the problem which would lead us too far afield but content ourselves with some sample results.

The historical development of non-commutative ring theory over the last 100 years is fascinating, and there are good accounts on it. We have therefore confined ourselves to historical remarks on three major players, Artin, Noether and Wedderburn, contained in the appropriate chapters. Each chapter is completed by a section of exercises, of varying difficulty. The vast majority of them merely test the understanding of the concepts introduced prior and are meant for the reader to get their hands on something concrete, and not as a serious challenge. In category theory proper one needs to work with 'classes' which we left undefined (as is common practice in basic texts, just think of a class as a set which may not be contained in another). A keen reader will find possible axiomatic approaches to category theory in the appendix of [30] or in [25], but there are others.

Finally, I would like to thank all my students who diligently attended my lectures and tried to learn what I had to offer. Several of them pointed out oversights in the online notes, and I am particularly indebted to my former students Linda Mawhinney and Michael Rosbotham for useful discussions on Chaps. 9 and 10, respectively.

Motivation from Ring Theory

In this chapter, the reader will become familiar with our notation and basic assumptions on the one hand; on the other, they are invited to various instances where the setting of Ring Theory in its strictest sense turns out to be too narrow, or at least somewhat cumbersome, in order to provide us with enough tools for studying certain properties of rings. We shall take this as part of our motivation to introduce the more comprehensive setting of modules below.

While the first section will be more informal in that we do not provide every detail of each argument, the second section will devoted to a few first steps in Category Theory, where we will introduce the fundamental concepts and illustrate these with a number of examples.

1.1 Basics on Modules

We begin this section with a reminder on the concept of a ring.

Definition 1.1 A *ring R* consists of a non-empty set together with two binary operations "+" and "·", which are usually referred to as the addition and the multiplication in R, such that

(i) $(R, +)$ is an abelian group;
(ii) $(R, \cdot)$ is a semigroup (which simply means · is associative);
(iii) the two distributivity laws

$$(x + y)z = xz + yz \quad \text{and} \quad z(x + y) = zx + zy \qquad (x, y, z \in R)$$

hold.

The ring R is called *unital* if $(R, \cdot)$ has an identity element, i.e., an element 1 such that $1 \cdot x = x = x \cdot 1$ for all $x \in R$. The ring R is called *commutative* if $(R, \cdot)$ is commutative.

Note that we will not automatically assume that every ring is unital but this assumption will be made explicit whenever necessary; the theory of unital rings runs more smoothly in general. We list a number of examples that are useful to understand the general theory.

Examples 1.2

(i) $\mathbb{Z}$, the ring of integers, is the most basic but still an interesting ring. Clearly it is both unital and commutative. Any new concept introduced should be checked for $\mathbb{Z}$ first.

(ii) $\mathbb{Z}^n = \mathbb{Z} \times \ldots \times \mathbb{Z}$, for a natural number n, with coordinate-wise operations. This gives another unital commutative ring which already has some different features. E.g., it is no longer an integral domain.

(iii) Every field K—such as $\mathbb{Q}$, $\mathbb{R}$ and $\mathbb{C}$—provides us with an extremely well-behaved commutative unital ring. To some extent, Ring Theory is about how far one can relax the nice multiplicative properties in a field and still get useful and interesting examples.

(iv) $\mathbb{H}$, the quaternions, almost forms a field but for the non-commutativity of multiplication. Thus, they yield the first interesting example of a non-trivial division ring. (See, for example, [34, Sect. 5.2].)

(v) $R[x]$, the ring of polynomials in the indeterminate x over a commutative unital ring R. When iterated, this construction yields the ring $R[x_1, \ldots, x_n], n \in \mathbb{N}$ of polynomials in n indeterminates which is highly useful in Algebraic Geometry.

(vi) $M_n(R)$, where R is a unital ring and $n \in \mathbb{N}$, is the ring of $n \times n$ matrices with entries in R. Even if R is commutative, this ring is only commutative in case $n = 1$. In general, the matrix rings are a source of useful examples for many ring theoretic properties. E.g., $M_n(R)$ is simple (that is, has no non-trivial two-sided ideals) if and only if R is simple. (See Proposition 7.1.1 in Chap. 7.)

(vii) $T_n(R)$, for a unital ring R and $n > 1$, is the ring of strictly upper diagonal matrices, that is, an $n \times n$ matrix (a_{ij}) belongs to $T_n(R)$ if $a_{ij} = 0$ for all $i \geq j$ and $a_{ij} \in R$ is arbitrary otherwise. This is a non-unital noncommutative ring.

In the above examples, the operations of addition and multiplication are the canonical ones; hence there is no need to discuss these in further detail as they are treated in every undergraduate textbook on ring theory such as [7] or [34], for instance. The next two important examples may be less familiar; for that reason, we spend a little more time on them. The first is fundamental for the development of module theory.

(viii) End(G), the endomorphism ring of an abelian group G. An *endomorphism* of an abelian group G is a mapping $\varphi \colon G \to G$ such that $\varphi(g+h) = \varphi(g)+\varphi(h)$

for all $g, h \in G$. (We write abelian groups additively, as is customary.) It follows that $\varphi(0) = 0$ and $\varphi(-g) = -\varphi(g)$. Composing two endomorphisms φ and ψ yields a new one $\psi \circ \varphi$ which gives the multiplication in $\text{End}(G)$. The addition is defined via $(\varphi + \psi)(g) = \varphi(g) + \psi(g)$, $g \in G$, and it is straightforward to verify the distributivity laws. The identity in $\text{End}(G)$ is of course the identity mapping, and in general, $\text{End}(G)$ is noncommutative.

(ix) $R[G]$, the group ring of an arbitrary group G. The elements of $R[G]$ are finite formal sums of the form $\sum_{g \in G} r_g g$ where $r_g \in R$ and only finitely many r_g are non-zero. The addition is defined via

$$\sum_{g \in G} r_g g + \sum_{g \in G} s_g g = \sum_{g \in G} (r_g + s_g)g$$

and the multiplication is given via convolution, that is,

$$\left(\sum_{g \in G} r_g g\right) \cdot \left(\sum_{h \in G} s_h h\right) = \sum_{k \in G} t_k k,$$

with $t_k = \sum_{k=gh} r_g s_h$. (Here, we write the group multiplicatively as it does not have to be abelian.) If R is unital then G is canonically a subgroup of $R[G]$ via $g \mapsto 1 \cdot g$, and if $e \in G$ is the identity then $1 \cdot e$ is the identity in $R[G]$. If both R and G are commutative, so is $R[G]$. In the case when K is a field, the group ring $K[G]$ in addition is a K-vector space with basis G (and the scalar multiplication simply defined via multiplication of the coefficients in K), so it is a K-algebra.

In this way, methods from Ring Theory can be used to study groups. For instance, one of the deep open and longstanding problems is whether a group G is torsion free if and only if the group ring $K[G]$ has no non-trivial divisors of zero. See also Exercise 1.3.6. (Quite a lot on this problem can be found in [31].) We shall discuss certain aspects of group rings in Chap. 10.

The aim of Ring Theory is to understand the structure and the properties of these and many other classes of rings. An important tool to achieve this aim is the concept of a *module* which allows us to look at a given ring from various perspectives.

Definition 1.3 Let R be a ring. Let M be an abelian group. We say that M is a *left R-module*, and denote this by writing $_R M$, if there is a bi-additive mapping $R \times M \to M$, $(r, m) \mapsto r \cdot m$ such that $(rs) \cdot m = r \cdot (s \cdot m)$ for all $r, s \in R$ and all $m \in M$. Here, *bi-additive* means

$$(r + s) \cdot m = r \cdot m + s \cdot m, \quad r \cdot (m + n) = r \cdot m + r \cdot n \quad (r, s \in R, \ m, n \in M).$$

In the case when R is unital and $1 \cdot m = m$ for each $m \in M$, the left module $_R M$ is called *unital* (in the literature, *unitary* is used as well).

There is an obvious symmetric way to define a (unital) *right R-module M_R*. If both sides work together we obtain the notion of a bimodule.

Definition 1.4 Let R and S be rings, and let M be an abelian group. We say that M is an *R-S-bimodule*, written as $_R M_S$, if M is a left R-module as well as a right S-module and the compatibility condition $r \cdot (m \cdot s) = (r \cdot m) \cdot s$ holds for all $r \in R$, $s \in S$ and $m \in M$.

Remark Let $_R M$ be a left R-module. Sometimes the mapping $(r, m) \mapsto r \cdot m$ is referred to as the "scalar multiplication" and the elements in R as the "scalars". This mapping is also called the "action" of R on $_R M$.

If one does not want to specify which kind of (one-sided) module one speaks of, one says "a module M over the ring R". We next come to the important concept of a substructure.

Definition 1.5 Let $_R M$ be a left R-module. Let N be a subgroup of M which is closed under the R-action, that is, $r \cdot n \in N$ for every $r \in R$ and $n \in N$. Then N is called a *left R-submodule* of $_R M$; we denote this fact by writing $_R N \leq _R M$.

Similarly, $N_R \leq M_R$ and $_R N_S \leq _R M_S$ are defined.

The submodules of a module are viewed as the (smaller) building blocks; the connections between different modules are given by the structure preserving mappings defined below.

Definition 1.6 Let $_R M$ and $_R N$ be left R-modules. A mapping $f : {}_R M \to {}_R N$ is called a *left R-module map* (also: a *left R-module homomorphism*) if it is additive, that is, $f(m_1 + m_2) = f(m_1) + f(m_2)$ for all $m_1, m_2 \in M$, and $r \cdot f(m) = f(r \cdot m)$ for all $r \in R$ and $m \in M$.

In a similar fashion, *right module maps* and *bimodule maps* between right modules and bimodules, respectively are defined.

We have now reached a point in our discussion where it is convenient to introduce a notation for the collection of all modules over a given ring. In the second part of this chapter, we shall, however, realise that this is much more than a mere shorthand.

Notation In order to save writing, we shall denote the collection of all left R-modules over a fixed ring R by $R\text{-}\mathcal{M}od$. Analogous meaning is given to the symbols $\mathcal{M}od\text{-}R$ and $R\text{-}\mathcal{M}od\text{-}S$ (where S is another ring). If R (and S) is unital, this notation will *automatically* refer to the collection of all *unital* modules.

Since the module maps are of paramount importance in the theory, we also give the sets comprising all of them special symbols.

Notation Let $_RM$, $_RN \in R\text{-}\mathcal{M}\!od$. Then

$$\text{Hom}_R \, (_RM, \, _RN) = \{f : \, _RM \to \, _RN \mid f \text{ is a left } R\text{-module map}\}.$$

Analogous meanings are given to $\text{Hom}_R \, (M_R, \, N_R)$ and $\text{Hom}_{R\text{-}S} \, (_RM_S, \, _RN_S)$.

Rather than delving further into the general theory of modules, let us now pause to ask the question "Why study modules at all?" The following instances of the appearance of modules shall provide us with motivation to develop a comprehensive theory as well as with some first examples.

1.1.1 Reducing Complicated Rings to Simpler Ones

Suppose R is a commutative unital ring. In the case when R is a field, every multiplicative equation $ax = b$, $a \neq 0$ can be solved uniquely. If R is merely an integral domain, at least every equation of the form $ax = 0$, $a \neq 0$ has the unique solution $x = 0$. This makes computations much easier: "Non-zero common factors can be cancelled." Unfortunately, the property of being an integral domain is all but too easily destroyed (Example 1.2 (ii) above). Is there hope that this trivial way of taking direct products might be the only method to destroy this good property of a ring? For certain classes of rings this is indeed (almost) true. A natural assumption, however, must be that there are no non-trivial nilpotent elements (which means that, if $x^n = 0$ for some $n \in \mathbb{N}$, then $x = 0$). Rings satisfying this property are called *reduced*.

The following is a well-known result in commutative Ring Theory; compare also [24, Sect. 12].

Theorem *Every commutative unital reduced ring is a subdirect product of integral domains.*

Let us indicate the argument for the sake of illustration. If R is such a ring then there is an injective ring homomorphism φ from R into the direct product $\prod_\alpha R_\alpha$ of integral domains R_α (with a possibly large index set of α's) such that $R_\alpha = \pi_\alpha \circ \varphi(R)$ for each α, where π_α is the canonical projection onto R_α. The key to this result is that, for every $x \in R \setminus \{0\}$, one can find a proper prime ideal P_x of R with the property $x \notin P_x$. The integral domain R_α is then defined by R/P_x (so that $\alpha = x$ and the index set is all of R !). See also Exercise 1.3.11.

The above method of taking appropriate homomorphic images and then embed the ring into a product of those is already very much in the spirit of module theory: we will see that quotient rings have a canonical module structure right away.

1.1.2   One-Sided Ideals in Noncommutative Rings

Suppose R is a noncommutative ring. One-sided ideals play an important role in this setting as R easily may have no non-trivial two-sided ideals (see Example 1.2 (vi) above). Say, L is a left ideal of R. Unfortunately the quotient group R/L does not carry a well-defined multiplication so it is not a ring. Nevertheless we can define a left R-action on R/L via

$$r \cdot (a + L) = ra + L \qquad (r, a \in R).$$

In this way, R/L turns into a left R-module which is unital if R is unital. Note also that $L \in R\text{-}\mathcal{M}od$ too!

We shall see that these two types of modules are very popular.

1.1.3   Images of Ideals Under Homomorphisms

Suppose R and S are rings, and $\rho\colon R \to S$ is a ring homomorphism. Let I be an ideal in R. (It does not matter here whether it is one-sided or two-sided so suppose that I is at least a left ideal.) In general, $\rho(I)$ will not be an ideal of S. However, if we consider R as a left module $_R R$ in the obvious way (the left action is just the ordinary multiplication), which is called the *standard R-module*, and we regard S as a left R-module via $r \cdot s = \rho(r)s$, $r \in R$, $s \in S$, then $\rho(I)$ becomes a left R-submodule of $_R S$. In fact, $\rho(I)$ is the image under the associated left R-module map ρ.

We learn from this observation that modules allow for a much larger flexibility.

1.1.4   Embedding into the Endomorphism Ring and General Representations

Every unital ring can be embedded into an endomorphism ring; thus, the latter may be considered as "the mother of all rings". Using the language of modules, this process, reviewed below, becomes very natural.

Let R be a unital ring. We now formalise the construction of the standard R-module $_R R$ indicated in the previous section. The abelian group $(R, +)$ becomes a unital left R-module in a natural way via

$$r \in R, \; s \in R\colon \quad r \cdot s = rs.$$

The endomorphism ring $\operatorname{End}((R, +))$ contains R as a unital subring in a canonical way:

$$\lambda\colon R \to \operatorname{End}((R, +)), \quad r \mapsto \lambda_r,$$

where $\lambda_r(s) = rs$, $s \in R$; this is the *left regular representation of R*. It is easy to check that λ is an injective unital ring homomorphism.

This representation of R corresponds precisely to the unital left R-module $_RR$. More generally, a *representation* of R is a pair (π, G) consisting of an abelian group G and a unital ring homomorphism $\pi : R \to \mathrm{End}(G)$. In this situation, we can turn G into a unital left R-module by defining

$$r \in R, \ g \in G: \quad r \cdot g = \pi(r)(g).$$

Conversely, suppose that $_RM \in R\text{-}\mathcal{M}od$; then

$$\pi_M : R \to \mathrm{End}(M), \quad \pi_M(r)(m) = r \cdot m \qquad (r \in R, \ m \in M)$$

defines a representation of R on G. It is easily verified that these two procedures are inverses to each other. Thus we have our first result.

Proposition 1.7 *Let R be a unital ring. There is a one-to-one correspondence between unital left R-modules and representations of R.*

1.1.5 Group Representations

Let G be a group, written multiplicatively. Let K be a field and $n \subset \mathbb{N}$ a natural number. By $\mathrm{GL}_n(K)$ we denote the group of invertible $n \times n$ matrices over K. Suppose $d : G \to \mathrm{GL}_n(K)$ is a representation of G, that is, a group homomorphism into $\mathrm{GL}_n(K)$. Put $R = K[G]$, the group ring of G (see Example 1.2 (ix) above). We can endow the abelian group K^n with a module structure as follows

$$\left(\sum_{g \in G} a_g g \right) \cdot v = \sum_{g \in G} a_g \, d(g)(v), \quad v \in K^n.$$

In this way, $K^n \in K[G]\text{-}\mathcal{M}od$.

Conversely, suppose $V \in K[G]\text{-}\mathcal{M}od$ is given. Since $K \subseteq K[G]$ as a unital subring via $a \mapsto ae$, V is a left K-module too, i.e., a K-vector space. Suppose that $\dim_K V = n \in \mathbb{N}$. As explained above in Sect. 1.1.4, this amounts to a representation $\rho : K[G] \to \mathrm{End}(V)$ such that $g \cdot v = \rho(g)(v)$ for $g \in G$, $v \in V$. By choosing a basis of V, we can identify $\mathrm{End}(V)$ with $M_n(K)$ and obtain in this way a representation $d : G \to \mathrm{GL}_n(K)$, $d(g) = \rho(g)$. A different choice of basis results in a representation $d' : G \to \mathrm{GL}_n(K)$ of G which is equivalent to d in the sense that $d'(g) = S^{-1}d(g)S$, $g \in G$, where $S : V \to V$ is the invertible transformation implementing the change of basis.

On the other hand, one can show that any pair of equivalent representations of G gives rise to isomorphic left $K[G]$-modules.

The above discussion shows that modules appear in many guises; therefore it should be fruitful to have a well-developed theory at hand. To complete this section, we record some (further) basic examples of modules.

Examples 1.8

(a) Every abelian group G becomes a $\mathbb{Z}$-module in a natural way: for $n \in \mathbb{N}$, $g \in G$ we define $n \cdot g = g + \ldots + g$, the sum of n copies of g. Furthermore, $0 \cdot g = 0_G$ and $-n \cdot g = -(n \cdot g)$. It is easy to verify the axioms of a unital $\mathbb{Z}$-module. Therefore, the theory of modules is a generalisation of the theory of abelian groups.

(b) Let V be a vector space over the field K. Then V is a K-module via:

$$r \in K, \ v \in V: \quad r \cdot v = rv \text{ (the usual scalar multiplication)}.$$

Therefore, the theory of modules is also a generalisation of the theory of vector spaces.

(c) Let V be a K-vector space with $\dim_K V = n < \infty$. Choose a basis for V and put $R = M_n(K)$. Then $V \in R\text{-}\mathcal{M}od$ via:

$$a \in M_n(K), \ v \in V: \quad a \cdot v \text{ is the usual matrix-vector multiplication}.$$

(d) Let V be a K-vector space, and let $T: V \rightarrow V$ be a linear mapping. Put $R = K[x]$. For $p(x) \in R$, the linear mapping $p(T)$ is well defined. Hence, V becomes a left R-module via

$$p(x) \in R, \ v \in V \quad p(x) \cdot v = p(T)(v).$$

To summarise this first section, we have encountered a number of instances where modules over rings appear in a rather natural fashion. We will now take a look at what we have achieved from a bird's eye view.

1.2   Categories of Modules

It is now time to start introducing the language of categories. Category Theory itself is nowadays a vast and highly developed area in Pure Mathematics with manifold applications in other fields. We will only be able to touch upon the most basic concepts and features; nevertheless this shall allow us a new view on the interplay between rings and modules.

S. Eilenberg and S. MacLane established the foundations in 1945 in their article "General theory of natural equivalences", see [15]. Following their conviction that

"It should be observed first that the whole concept of a category is essentially an auxiliary one; our basic concepts are essentially those of a functor and of a natural transformation."

we will very soon include functors (as well as natural transformations) in our discussion; see Sect. 3.2.

In the following, we will have to deal with "very large collections of sets" which will be called *classes*. The reader familiar with Russell's Paradox will know that we cannot perform arbitrary operations with sets without creating set-theoretic problems. The solution to this situation is a careful distinction between actual sets and more flexible classes; this however needs an axiomatic approach to Set Theory which would go far beyond the scope of this book. Theories which extend the Zermelo–Fraenkel set theory together with the Axiom of Choice and allow a rigorous treatment are the von Neumann–Bernays–Gödel axioms or the Morse–Kelley set theory. An alternative is to work within Grothendieck universes. The interested reader will find answers to their questions in [30], for example. The main difference between sets and classes is that any class that is an element of another class is a set.

Definition 1.9 A *category* $\mathscr{C}$ is a triple consisting of an object class ob($\mathscr{C}$); a morphism class mor($\mathscr{C}$); and a law of composition $\circ$; that is,

(i) ob($\mathscr{C}$) is a class, its elements $A, B, C, \ldots$ are called *the objects of* $\mathscr{C}$;
(ii) mor($\mathscr{C}$) is a class, its elements are pairwise disjoint sets Mor(A, B), for $A, B \in$ ob($\mathscr{C}$); each element in Mor(A, B) is called a *morphism* or an *arrow* from A to B; typically we write these as $f, g, h, \ldots$;
(iii) $\circ$ yields a family of mappings

$$\text{Mor}(A, B) \times \text{Mor}(B, C) \to \text{Mor}(A, C), \quad (f, g) \mapsto g \circ f,$$

where $A, B, C \in$ ob($\mathscr{C}$), such that the following two axioms hold
(a) (associativity)

$$\forall f \in \text{Mor}(A, B), \ g \in \text{Mor}(B, C), \ h \in \text{Mor}(C, D) :$$

$$h \circ (g \circ f) = (h \circ g) \circ f;$$

(b) (identities)

$$\forall A \in \text{ob}(\mathscr{C}) \ \exists_1 \ 1_A \in \text{Mor}(A, A) \text{ such that}$$

$$\forall B \in \text{ob}(\mathscr{C}) \ \forall f \in \text{Mor}(A, B), \ g \in \text{Mor}(B, A) :$$

$$f = f \circ 1_A \quad \text{and} \quad 1_A \circ g = g.$$

1_A is called the *identity morphism* of A.

The category $\mathscr{C}$ is called *concrete* if the elements of ob($\mathscr{C}$) have underlying sets and the morphisms of mor($\mathscr{C}$) are mappings between these underlying sets. (A more precise definition will be given in 3.2.2.) The category $\mathscr{C}$ is called *small* if $\mathrm{ob}(\mathscr{C})$ is a set.

Notation Suppose $f \in \mathrm{Mor}(A, B)$ for two objects A, B in a category $\mathscr{C}$. We say A is *the domain of* f, and write $A = \mathrm{dom}(f)$, and we say B is *the codomain of* f, and write $B = \mathrm{codom}(f)$ in this case. If there is a need to specify the category certain morphisms belong to, we emphasise this by writing $\mathrm{Mor}_{\mathscr{C}}(A, B)$, $A, B \in \mathscr{C}$.

In order to simplify the notation, we shall sometimes write $A \in \mathscr{C}$ instead of $A \in \mathrm{ob}(\mathscr{C})$ and $f \in \mathscr{C}$ instead of $f \in \mathrm{Mor}(A, B) \in \mathrm{mor}(\mathscr{C})$ though $f \in \mathrm{Mor}(A, B)$ will be the most common.

Let us have a look at some of the categories the reader may have encountered.

1.10 Examples of Categories

$\mathscr{S}$	the category of all sets where the objects are sets and the morphisms are mappings between sets.
$\mathscr{Top}$	the category of all topological spaces; the objects are topological spaces and the morphisms are continuous mappings between them.
$\mathscr{Comp}$	the category of compact Hausdorff spaces; the morphisms are again the continuous mappings.
$\mathscr{Gr}$	the category of groups with group homomorphisms as morphisms.
$\mathscr{AGr}$	the category of abelian groups with group homomorphisms as morphisms.
$\mathscr{Ring}$	the category of rings with ring homomorphisms as morphisms.
$\mathscr{Ring}_1$	the category of unital rings with unital ring homomorphisms as morphisms.
$R\text{-}\mathscr{Mod}$	the category of (unital) left R-modules over a given (unital) ring R where the morphisms are the left R-module maps.
$\mathscr{Ban}_\infty$	the category of complex Banach spaces with bounded linear operators as the morphisms.
$\mathscr{Ban}_1$	the category of complex Banach spaces with linear contractions as the morphisms.
$\mathscr{Lat}$	the category of lattices and lattice homomorphisms as the morphisms.
$\mathscr{HTop}$	the category of all topological spaces and homotopy classes of continuous mappings as the morphisms.

All of the above categories but for the last one are concrete categories; the objects have an underlying set structure and the morphisms arise from mappings between these sets. Moreover, the composition law is provided by the composition of mappings. In $\mathscr{HTop}$, this is not the case.

The reader will have no difficulties in defining the categories $\mathscr{Mod}\text{-}R$ and $R\text{-}\mathscr{Mod}\text{-}S$ for themselves.

Although many of the above categories look alike on the surface, they have rather different properties, as we shall see when moving on in the later chapters. For example, for two objects $_RM, _RN \in R\text{-}\mathcal{M}od$ the set of morphisms $\mathrm{Mor}(_RM, _RN) = \mathrm{Hom}_R(_RM, _RN)$ has a group structure (see Exercise 1.3.2) whereas this is not the case for $X, Y \in \mathrm{ob}(\mathcal{T}op)$: there is no canonical addition for continuous mappings between X and Y.

Definition 1.11 A *subcategory* $\mathcal{D}$ of a category $\mathcal{C}$ consists of a subclass $\mathrm{ob}(\mathcal{D}) \subseteq \mathrm{ob}(\mathcal{C})$, a subclass $\mathrm{mor}(\mathcal{D}) \subseteq \mathrm{mor}(\mathcal{C})$ in the sense that $\mathrm{Mor}(A, B) \in \mathrm{mor}(\mathcal{D})$ if $A, B \in \mathrm{ob}(\mathcal{D})$, and $\circ$ is simply the restriction of the composition of morphisms in $\mathcal{C}$ to $\mathcal{D}$.

The subcategory $\mathcal{D}$ is said to be *full* if, for all $A, B \in \mathrm{ob}(\mathcal{D})$, every morphism $A \to B$ in $\mathcal{C}$ is also a morphism in $\mathcal{D}$, that is, $\mathrm{Mor}_{\mathcal{D}}(A, B) = \mathrm{Mor}_{\mathcal{C}}(A, B)$.

For instance, $\mathcal{C}omp$ is a full subcategory of $\mathcal{T}op$ and $\mathcal{A}\mathcal{G}r$ is a full subcategory of $\mathcal{G}r$. On the other hand, $\mathcal{R}ing_1$ is a non-full subcategory of $\mathcal{R}ing$ and $\mathcal{B}an_1$ is a non-full subcategory of $\mathcal{B}an_\infty$. Note that $\mathcal{R}ing$ is not even a subcategory of $\mathcal{A}\mathcal{G}r$ as an abelian group can carry several non-isomorphic rings structures.

Let $R\text{-}mod$ denote the category whose objects are the finitely generated left modules over a fixed ring R (i.e., those modules that have a finite set of generators, see Definition 2.3.1) together with the left R-module maps. Then $R\text{-}mod$ is a full subcategory of $R\text{-}\mathcal{M}od$.

Remark 1.12 Even if the objects in a category have an underlying set (and are specified by some additional structure on these sets), the morphisms do not have to be mappings on the underlying sets. An example is $\mathcal{H}\mathcal{T}op$. Another one arises from the open subsets of a given topological space X: the category $\mathcal{O}_X$ has as objects all open subsets of X and, for $U, V \in \mathrm{ob}(\mathcal{O}_X)$, $\mathrm{Mor}(U, V) = \{\to\}$ if $U \subseteq V$ and $\mathrm{Mor}(U, V) = \emptyset$ otherwise. The composition law is defined through the transitivity of set inclusion.

Let us have a look at some small categories.

1.13 Examples of Small Categories

(i) Let X be a non-empty set, and let $\sim$ be an equivalence relation on X. Define a category $\mathcal{E}q_X$ by taking as the objects in $\mathcal{E}q_X$ the elements of X and for $x, y \in X$ set

$$\mathrm{Mor}(x, y) = \begin{cases} \{(x, y)\} & \text{if } x \sim y \\ \emptyset & \text{otherwise.} \end{cases}$$

The transitivity of $\sim$ yields the composition law $\circ$.

(ii) Let G be a group. We can consider G as a category whose single object is G itself and whose morphism class (which is a set in this case) is G too, that is, $\mathrm{Mor}(G, G) = G$ in the following sense: we identify $g \in G$ with the morphism $h \mapsto gh$, and the composition law is provided by the group operation.

Some of the morphisms in a category are of special importance; we shall introduce these now.

Definition 1.14 Let $\mathscr{C}$ be a category. Let $A, B \in \mathrm{ob}(\mathscr{C})$. A morphism $f \in \mathrm{Mor}(A, B)$ is said to be a *monomorphism* if, for each $C \in \mathrm{ob}(\mathscr{C})$ and all $g, h \in \mathrm{Mor}(C, A)$, $fg = fh$ implies $g = h$. The morphism f is said to be an *epimorphism* if, for each $C \in \mathrm{ob}(\mathscr{C})$ and all $g, h \in \mathrm{Mor}(B, C)$, $gf = hf$ implies $g = h$. (We wrote gf for $g \circ f$, etc. as is customary.) The morphism f is called an *isomorphism* if there exists $g \in \mathrm{Mor}(B, A)$ such that $gf = 1_A$ and $fg = 1_B$. (In this case, g is unique.) The objects A and B are called *isomorphic* if there exists a isomorphism between them.

We conclude this section with a basic but very useful construction.

Example 1.15 Let $\mathscr{C}$ be a category. The *dual category* $\mathscr{C}^{\mathrm{op}}$ is defined by $\mathrm{ob}(\mathscr{C}^{\mathrm{op}}) = \mathrm{ob}(\mathscr{C})$ and for $A, B \in \mathrm{ob}(\mathscr{C}^{\mathrm{op}})$ we put $\mathrm{Mor}_{\mathscr{C}^{\mathrm{op}}}(A, B) = \mathrm{Mor}_{\mathscr{C}}(B, A)$, that is, we "reverse" all arrows and the composition in $\mathscr{C}^{\mathrm{op}}$ is the same as in $\mathscr{C}$. This dualisation procedure has many nice features; e.g., a monomorphism in $\mathscr{C}$ becomes an epimorphism in the dual category.

1.3   Exercises

Exercise 1.3.1 Let $_R M$ and $_R N$ be in $R\text{-}\mathscr{M}od$. For a left R-module map $f : {}_R M \to {}_R N$, we define its *kernel* by

$$\ker(f) = \{ m \in {}_R M \mid f(m) = 0 \}$$

and its *image* by

$$\mathrm{im}(f) = \{ f(m) \mid m \in {}_R M \}.$$

Show that $\ker(f)$ is a left R-submodule of $_R M$ and $\mathrm{im}(f)$ is a left R-submodule of $_R N$.

Exercise 1.3.2 For two module maps f and g defined on a module M with values in a module N over the ring R define their sum by $(f + g)(m) = f(m) + g(m)$ for all $m \in M$. Show that $\mathrm{Hom}_R({}_R M, {}_R N)$ and $\mathrm{Hom}_{R\text{-}S}({}_R M_S, {}_R N_S)$ (where both M and N are R-S-bimodules and S is another ring) are abelian groups under this addition.

Exercise 1.3.3 Let $\rho : R \to S$ be a ring homomorphism between the two rings R and S. Let $_S M \in S\text{-}\mathscr{M}od$. Show that M becomes a left R-module under the action $r \cdot m = \rho(r) \cdot m$, $r \in R$ and $m \in M$.

Exercise 1.3.4 Let R be a ring without identity. Show that we can embed R into a unital ring in the following way: equip the abelian group $R \times \mathbb{Z}$ with the multiplication

$$(r, n) \cdot (s, m) = (rs + ns + mr, nm) \qquad (r, s \in R, \ n, m \in \mathbb{Z}).$$

Verify that we indeed get a ring multiplication in this way, and that $(0, 1)$ serves as an identity. The ring thus obtained is denoted by $R^{\times}$ and is called *the minimal unitisation of R* or the *Dorroh superring of R*.

Show further that the monomorphism $r \mapsto (r, 0)$ allows us to view R as an ideal in $R^{\times}$. (However, if performed for a ring which already has an identity, this process will destroy the original identity!)

Exercise 1.3.5 Suppose R is a unital ring and that $_R M$ is a left R-module. Then

$$_R M_0 = \{m \in M \mid r \cdot m = 0 \ \forall r \in R\} \ \text{ and } \ _R M_1 = \{m \in M \mid 1 \cdot m = m\}$$

are both submodules of $_R M$. Moreover, $_R M = {_R M_0} \oplus {_R M_1}$. Prove these statements!

The submodule $_R M_1$ is unital and the submodule $_R M_0$ carries only the trivial R-action; so all information on $_R M$ is actually contained in $_R M_1$.

Exercise 1.3.6 Let G be a group with non-trivial torsion; that is, there is $g \in G$ such that $g^n = e$ for some $n \in \mathbb{N}$ and $g \neq e$, where e is the neutral element in G. Show that the group ring $R[G]$ contains a non-trivial divisor of zero.

Exercise 1.3.7 Let H be a subgroup of a group G. Show that the mapping $\tau_H \colon R[G] \to R[H]$ defined by $\tau_H\left(\sum_{g \in G} r_g g\right) = \sum_{g \in H} r_g g$ is a $R[H]$-module map. (Use the support of $a \in R[G]$ defined by $\mathrm{supp}(a) = \{g \in G \mid r_g \neq 0\}$, where $a = \sum_{g \in G} r_g g$.)

Exercise 1.3.8 Let H be a subgroup of a group G. Let $a \in R[H]$. Use the previous exercise to show that, if a is invertible in $R[G]$, then it is invertible in $R[H]$, and that, if a is a zero divisor in $R[G]$, then it is a zero divisor in $R[H]$.

Exercise 1.3.9 Let G be a group and R be a unital ring. Define the *trace* on $R[G]$ by $\mathrm{tr}\left(\sum_{g \in G} r_g g\right) = r_1$, the coefficient at the identity $1 \in G$. Show that tr is a left R-module map and that $\mathrm{tr}(ab) = \mathrm{tr}(ba)$ for all $a, b \in R[G]$.

Exercise 1.3.10 Let R be a commutative unital ring, and denote by $\mathrm{nil}(R)$ the set of all nilpotent elements in R. Show that $\mathrm{nil}(R)$ is an ideal of R; it is called the *nil radical* of R. Show further that the quotient ring $R/\mathrm{nil}(R)$ does not contain any non-zero nilpotent element.

Exercise 1.3.11 Let R be a unital reduced ring. A proper ideal P of R is said to be *prime* if, whenever I_1, I_2 are ideals of R such that $I_1 I_2 \subseteq P$, we have $I_1 \subseteq P$ or

$I_2 \subseteq P$. Using Zorn's Lemma, show that the intersection of all prime ideals of R is zero.

Exercise 1.3.12 Let (π, G) be a representation of the unital ring R (that is, $\pi : R \to \text{End}(G)$ is a unital ring homomorphism from R into the endomorphism ring of the abelian group G). Show that the action of R on G defined by $r \cdot g = \pi(r)(g), r \in R$, $g \in G$ turns G into a left R-module.

Exercise 1.3.13 Let $\mathscr{C}$ be a concrete category. Show that a morphism in $\mathscr{C}$ which is an injective map is a monomorphism and a morphism which is a surjective map is an epimorphism. Does this imply that a bijective morphism is an isomorphism in $\mathscr{C}$?

Exercise 1.3.14 Let R be a unital ring. Show that every monomorphism in $R\text{-}\mathcal{M}od$ is injective and every epimorphism in $R\text{-}\mathcal{M}od$ is surjective. Does a similar result hold in the full subcategory $R\text{-}mod$ of $R\text{-}\mathcal{M}od$?

Exercise 1.3.15 Show that the canonical inclusion $\mathbb{Z} \hookrightarrow \mathbb{Q}$ is an epimorphism but not surjective.

Exercise 1.3.16 Let E and F be Banach spaces and let $f : E \to F$ be a morphism from E to F in $\mathscr{B}an_1$. Show that f is a monomorphism if and only if $f^{-1}(0) = 0$ and that f is an epimorphism if and only if $f(E)$ is dense in F. Are the same statements true for $f \in \text{mor}(\mathscr{B}an_\infty)$?

Constructions with Modules

2

This chapter is devoted to various basic constructions that we can perform within $R\text{-}\mathcal{M}od$. Once we have found some interesting modules, we want to be able to obtain new ones from them. Special classes of modules will play a distinguished role here. In the second part of this chapter we shall review these constructions from the viewpoint of category theory and thus realise that seemingly similar categories can behave rather differently.

2.1   Some Special Morphisms

Let R be a ring, and let $_R M$, $_R N \in R\text{-}\mathcal{M}od$. We say that $f \in \mathrm{Hom}_R\,(_R M, _R N)$ is

a *monomorphism* if f is injective;
an *epimorphism* if f is surjective;
an *isomorphism* if f is bijective.

The two modules $_R M$ and $_R N$ are called *isomorphic* if there is an R-isomorphism between them. Obvious modifications yield analogous notions for other types of modules.

It follows from Exercises 1.3.13 and 1.3.14 that the above definition is consistent with the one given in Definition 1.14.

Notation Let R be a ring, and let M be an R-module. Then

$$\mathrm{End}_R\,(_R M) = \mathrm{Hom}_R\,(_R M, _R M) \quad \text{and} \quad \mathrm{End}_R\,(M_R) = \mathrm{Hom}_R\,(M_R, M_R).$$

If S is another ring, then $\mathrm{End}_{R\text{-}S}\,(_R M_S) = \mathrm{Hom}_{R\text{-}S}\,(_R M_S, _R M_S).$

© The Author(s), under exclusive license to Springer Nature Switzerland AG 2022
M. Mathieu, *Classically Semisimple Rings*,
https://doi.org/10.1007/978-3-031-14209-3_2

In the last chapter, we already met submodules and homomorphic images of modules. The connection between the two is provided by the following fundamental construction.

2.2   Quotient Modules

Let R be a ring, and let $_RN$, $_RM \in R\text{-}\mathcal{M}od$ with $_RN \leq {}_RM$. On the (abelian) quotient group M/N we define an R-action by

$$r \in R, \ m \in M: \quad r \cdot (m + N) = r \cdot m + N.$$

This is well defined as, if $m + N = n + N$, then $m - n \in N$ and thus $r \cdot (m - n) \in N$. It follows that

$$r \cdot n + N = r \cdot m - r \cdot (m - n) + N = r \cdot n + N.$$

In this way we obtain the *quotient module* $_RM/_RN$; if R is unital, this construction gives the correct object in $R\text{-}\mathcal{M}od$. We also have the *canonical epimorphism*

$$\pi_N : {}_RM \to {}_RM/_RN, \quad \pi_N(m) = m + N.$$

Clearly, $\ker(\pi_N) = {}_RN$.

Proposition 2.2.1 *Let R be a ring, and let $_RN$, $_RM \in R\text{-}\mathcal{M}od$ with $_RN \leq {}_RM$. Then there is a one-to-one correspondence between the submodules of $_RM/_RN$ and the submodules of $_RM$ containing $_RN$.*

Proof It is easily checked that the inverse image of a submodule under a module map is a submodule of the domain. Therefore, for a submodule $_RL \leq {}_RM/_RN$, $_RK = \pi_N^{-1}(_RL) = \{m \in M \mid \pi_N(m) \in L\}$ is a submodule of $_RM$ containing $_RN$. Conversely, let $_RK \leq {}_RM$ be such that $_RN \subseteq {}_RK$. Then $\pi_N(_RK)$ is a submodule of $_RM/_RN$ (compare Exercise 1.3.1).

Evidently, $_RK \subseteq \pi_N^{-1}(\pi_N(_RK))$. If $m \in \pi_N^{-1}(\pi_N(_RK))$ then $\pi_N(m) \in \pi_N(_RK)$ and so m differs from an element of $_RK$ by an element in $_RN \subseteq {}_RK$. Consequently, $m \in {}_RK$ and we obtain $_RK = \pi_N^{-1}(\pi_N(_RK))$. On the other hand, we always have $\pi_N(\pi_N^{-1}(_RL)) \subseteq {}_RL$ for any $_RL \leq {}_RM/_RN$, and since π_N is surjective, we get the equality of the two sets. $\square$

Definition 2.2.2 Let R be a ring, and let $_RM \in R\text{-}\mathcal{M}od$. For a non-empty subset $S \subseteq {}_RM$ we let

$$\mathrm{Ann}_R(S) = \{r \in R \mid r \cdot m = 0 \text{ for all } m \in S\}$$

be the *annihilator of S in R*.

The module $_RM$ is said to be *faithful* if $\text{Ann}_R(_RM) = \{0\}$.

Clearly, $\text{Ann}_R(S)$ is always a left ideal of R and $\text{Ann}_R(_RM)$ is a two-sided ideal of R. It goes without saying that all of the above can be defined analogously for right modules and bimodules.

2.3 Generating Modules

In this section, we shall take a look at various ways to generate submodules from subsets of a given module.

Let $_RM \in R\text{-}\mathcal{Mod}$ for some ring R. Given a family $\{_RM_i \mid i \in I\}$ of submodules $_RM_i \leq {}_RM$, their intersection is easily seen to be a submodule of $_RM$. Suppose now that $S \subseteq {}_RM$ is a non-empty subset. The intersection of all submodules of $_RM$ that contain S—which is evidently the smallest submodule of $_RM$ containing S—is called *the submodule generated by S* and will be denoted by $_R\langle S \rangle$.

Suppose that R is unital. For $m \in M$, $_R\langle\{m\}\rangle = \{r \cdot m \mid r \in R\} = Rm$ and for any $\emptyset \neq S \subseteq {}_RM$, we have

$$_R\langle S \rangle = \sum_{s \in S} Rs = \left\{ \sum_{j \in J} r_j \cdot s_j \mid r_j \in R,\ s_j \in S,\ J \text{ finite} \right\}.$$

Definition 2.3.1 Let $_RM \in R\text{-}\mathcal{Mod}$ for a unital ring R. We say

(i) $_RM$ is *cyclic* if $M = Rm$ for some $m \in M$;
(ii) $_RM$ is *finitely generated* if $_RM = Rm_1 + \ldots + Rm_k$ for a finite subset $\{m_1, \ldots m_k\} \subseteq M$.

The elements m and $m_1, \ldots, m_k$ above are called *generators*.

2.4 Direct Sums and Products of Modules

Let $\{_RM_i \mid i \in I\}$ be a family of left R-modules. On the cartesian product $\underset{i \in I}{\times} {}_RM_i$ we define module operations as follows

$$(m_i) + (n_i) = (m_i + n_i),$$

$$r \cdot (m_i) = (r \cdot m_i)$$

for all $(m_i), (n_i) \in \underset{i \in I}{\times} {}_RM_i$ and $r \in R$. It is easy to check that we indeed obtain a left R-module which is denoted by $\underset{i \in I}{\prod} {}_RM_i$ and called *the direct product of the family* $\{_RM_i \mid i \in I\}$.

The submodule

$$\bigoplus_{i \in I} {}_R M_i = \left\{ (m_i) \in \prod_{i \in I} {}_R M_i \mid \text{ at most finitely many } m_i \text{ are non-zero} \right\}$$

is called *the direct sum of the family* $\{{}_R M_i \mid i \in I\}$.

There are some canonical module maps associated with these two modules:

$$p_j \colon \prod_{i \in I} {}_R M_i \to {}_R M_j, \ (m_i) \mapsto m_j \quad \text{for some } j \in I$$

is called the *canonical projection onto the j-th component*; it is an R-epimorphism.

$$\iota_j \colon {}_R M_j \to \bigoplus_{i \in I} {}_R M_i, \ m \mapsto (m_i) \text{ where } m_j = m \text{ and } m_i = 0 \text{ for } i \neq j$$

is called the *canonical injection from the j-th component*; it is an R-monomorphism.

The two constructions above are characterised by some universal properties.

Proposition 2.4.1 *Let* $\{{}_R M_i \mid i \in I\}$ *be a family of left R-modules.*

(i) *For every ${}_R N \in R$-$\mathcal{M}od$ and every family* $\{f_i \mid i \in I\} \subseteq \operatorname{Hom}_R ({}_R N, {}_R M_i)$ *there is a unique $f \in \operatorname{Hom}_R ({}_R N, \prod_i {}_R M_i)$ such that $p_j f = f_j$, $j \in I$.*

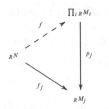

(ii) *For every ${}_R N \in R$-$\mathcal{M}od$ and every family* $\{f_i \mid i \in I\} \subseteq \operatorname{Hom}_R ({}_R M_i, {}_R N)$ *there is a unique $f \in \operatorname{Hom}_R (\bigoplus_i {}_R M_i, {}_R N)$ such that $f \iota_j = f_j$, $j \in I$.*

Proof

(i) For $n \in N$, we set $f(n)_i = f_i(n)$ for each $i \in I$. Then $p_i f = f_i$ and it is easy
   to verify that f is a left R-module map.
(ii) For $m \in \bigoplus_i {}_R M_i$, there are only finitely many i's with $m_i \neq 0$. Hence we can
   define $f(m) = \sum_i f_i(m_i)$ and obtain a left R-module map satisfying $f \iota_j = f_j$
   for all $j \in I$.

$\square$

2.5 Free Modules

This section deals with those modules that resemble vector spaces most closely. In
a way, they turn out to be the most general modules.

Let X be a non-empty set, and let R be a unital ring. We consider the set

$$\{ f \in R^X \mid f(x) \neq 0 \text{ for at most finitely many } x \in X \}$$

and write its elements as (finite sums) $\sum_{x \in X} r_x x$. We can regard X as a subset via
$x \mapsto 1 \cdot x$.

An R-module structure can be defined by

$$\sum_x r_x x + \sum_x s_x x = \sum_x (r_x + s_x) x,$$

$$r \cdot \sum_x r_x x = \sum_x r r_x x \qquad (r \in R).$$

In this way, we obtain a left R-module called *the free left R-module on X* which is
denoted by $R\langle X \rangle$ (sometimes also denoted as $R^{(X)}$). Note that, if we put $M_x = {}_R R$
for each $x \in X$, then $R\langle X \rangle$ is nothing but $\bigoplus_{x \in X} M_x$.

Clearly, analogous constructions give us free right modules and free bimodules,
respectively.

Free modules are distinguished in several ways.

Definition 2.5.1 Let ${}_R M \in R\text{-}\mathcal{M}od$. A subset $S \subseteq M$ is called *R-linearly inde-*
pendent (or simply, *linearly independent*) if, for every finite subset $\{ m_1, \ldots, m_k \} \subseteq$
S and elements $r_1, \ldots, r_k \in R$, $\sum_{i=1}^k r_i \cdot m_i = 0$ implies that $r_i = 0$ for all
$1 \leq i \leq k$. The set S is a *basis* of ${}_R M$ (or, more precisely, an *R-basis*) if it is
linearly independent and $R\langle S \rangle = {}_R M$.

Considering the elements of X inside the free module $R\langle X \rangle$, we see that X is a
basis of it.

Proposition 2.5.2 *Let R be a unital ring. Let $_R M \in R$-$\mathcal{M}od$ with basis S. Then $_R M$ is isomorphic to $R\langle S\rangle$ as left R-modules.*

Proof It is easy to check that the mapping $S \to R\langle S\rangle$, $s \mapsto 1 \cdot s$ extends to an isomorphism between $_R M$ and $R\langle S\rangle$. □

Remark Contrary to the situation of vector spaces over fields, the notion of the 'length of a basis' (and hence, the notion of 'dimension') is not well defined for free modules. For example, let $V = \mathbb{Q}[x]$ and $R = \mathrm{End}_{\mathbb{Q}}(_{\mathbb{Q}}V)$. It can be shown that $_R R \cong {}_R R \oplus {}_R R$ and hence $_R R^n \cong {}_R R^m$ for all $n, m \in \mathbb{N}$. If R is commutative, such a phenomenon cannot occur; see [8, Theorem 7.12].

Next we state the universal property of free modules.

Theorem 2.5.3 *Let R be a unital ring, and let $_R M \in R$-$\mathcal{M}od$. Let X be a non-empty set. If $\eta \colon X \to M$ is a mapping then there is a unique $f \in \mathrm{Hom}_R (R\langle X\rangle, {}_R M)$ such that $f(x) = \eta(x)$ for all $x \in X$.*

Proof We set $f\left(\sum_x r_x x\right) = \sum_x r_x \eta(x)$, $x \in X$. Evidently, f is a left R-module map extending the mapping $X \to R\langle X\rangle$, $x \mapsto 1 \cdot x$. Therefore $f(x) = \eta(x)$ for every $x \in X$. It is clear that f is uniquely determined by these properties. □

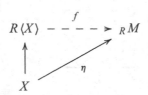

Corollary 2.5.4 *Every unital left module is a quotient of a free left module.*

Proof Let $_R M \in R$-$\mathcal{M}od$ for a unital ring R. Let $\eta \colon M \to M$ be the identity. By Theorem 2.5.3, there is a left R-module map $f \colon R\langle M\rangle \to {}_R M$ with $f(m) = m$ for every $m \in M$. Thus f is surjective. Let $_R N = \ker(f)$ which is a submodule of $R\langle M\rangle$ and define $\hat{f} \colon R\langle M\rangle /_R N \to {}_R M$ by $\hat{f}(a + {}_R N) = f(a)$, $a = \sum_m r_m m \in R\langle M\rangle$. Then $\hat{f}$ is an epimorphism and, since $f(a) = 0$ precisely when $a \in {}_R N$, it follows that $\hat{f}$ is injective too. As a result, $_R M$ is isomorphic to a quotient of a free module. □

We shall see in the next chapter that the technique used in the proof of the above corollary allows us to identify any homomorphic image of a module as a quotient module.

2.6   Special Objects in a Category

We now return to the viewpoint of categories. Motivated by the various constructions with modules that we performed in the previous sections of this chapter,
we want to investigate when similar objects in a category might be available.
Constructions such as the direct product, for example, can be done in many of the
categories listed in Sect. 1.2; however, they often depend on manipulations with the
underlying sets. In a general category, we thus have to identify properties of objects
that can be solely expressed in terms of morphisms—this is where the "universal
properties" come in—and the perspective turns from construction to the one of
existence.

Our first special objects continue the line of thought of the very last section.

2.6.1   Free Objects

Let $\mathscr{C}$ be a concrete category. Given a non-empty set X, an object $F \in \mathrm{ob}(\mathscr{C})$ and
an injective mapping $X \to F$ we say F *is free on* X if, for every object $A \in \mathrm{ob}(\mathscr{C})$
and every mapping $\eta \colon X \to A$, there is a unique $f \in \mathrm{Mor}_{\mathscr{C}}(F, A)$ such that $\eta(x) =
f(x)$ for all $x \in X$; that is, the diagram below is commutative.

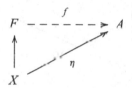

Note that we do not use a special symbol for the underlying set of an object in $\mathscr{C}$.

For example, for every unital ring R, if X is a non-empty set then $R\langle X \rangle$ is a free
object in $R\text{-}\mathscr{M}\!od$, by Theorem 2.5.3. On the other hand, the only free object in
$\mathbb{Z}\text{-}mod$ (the finitely generated $\mathbb{Z}$-modules) is the zero module.

2.6.2   Products and Coproducts

Products and direct sums (called "coproducts") are defined in a general category $\mathscr{C}$
just as in $R\text{-}\mathscr{M}\!od$.

Let $\{A_i \mid i \in I\}$ be a family of objects in a category $\mathscr{C}$. A *product* of the family
$\{A_i \mid i \in I\}$ consists of an object $A \in \mathrm{ob}(\mathscr{C})$ together with a family $\{p_i \mid i \in I\}$
of morphisms in $\mathrm{Mor}(A, A_i)$, $i \in I$, called *projections*, such that the following
diagram can be made commutative by a unique $f \in \mathrm{Mor}(C, A)$ for each $C \in \mathrm{ob}(\mathscr{C})$

and $f_i \in \mathrm{Mor}(C, A_i), i \in I$:

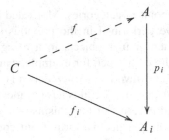

If such product exists, it is unique up to isomorphism and is denoted by $\prod_{i \in I} A_i$.

Dually, a *coproduct* of $\{A_i \mid i \in I\}$ consists of an object $A \in \mathrm{ob}(\mathscr{C})$ together with a family $\{e_i \mid i \in I\}$ of morphisms in $\mathrm{Mor}(A_i, A), i \in I$, called *injections*, such that the following diagram can be made commutative by a unique $g \in \mathrm{Mor}(A, B)$ for each $B \in \mathrm{ob}(\mathscr{C}), g_i \in \mathrm{Mor}(A_i, B), i \in I$:

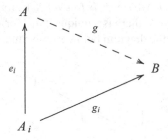

If such coproduct exists, it is unique up to isomorphism and is denoted by $\coprod_{i \in I} A_i$.

These two notions are defined in such a way that the following relations hold:

$$\prod_{i \in I} \mathrm{Mor}(C, A_i) \cong \mathrm{Mor}\left(C, \prod_{i \in I} A_i\right)$$

$$\prod_{i \in I} \mathrm{Mor}(A_i, B) \cong \mathrm{Mor}\left(\coprod_{i \in I} A_i, B\right).$$

$$(2.6.1)$$

Examples

(i) In the category $\mathscr{S}$, the cartesian product $\mathsf{X}_{i \in I} A_i$ of a family $\{A_i \mid i \in I\}$ of sets together with the projections onto the coordinates forms the product. The coproduct in $\mathscr{S}$ is the disjoint union $\coprod_{i \in I} A_i = \bigcup_{i \in I} (A_i \times \{i\})$ together with the injections $e_i : A_i \to \coprod_{i \in I} A_i, a \mapsto (a, i), a \in A_i$.

(ii) In $R\text{-}\mathscr{M}od$, the product is the direct product of a family of modules and the coproduct is the direct sum, together with the canonical projections and the canonical injections, cf. Proposition 2.4.1. In $R\text{-}mod$ however, there exist only finite products and coproducts (the index set of the family has to be finite).

(iii) In $\mathscr{B}an_1$, the direct product of a family $\{E_i \mid i \in I\}$ of complex Banach spaces is the Banach space of all bounded functions $(x_i)_{i \in I}$ from I into $\mathsf{X}_{i \in I} E_i$ endowed with the sup-norm; the projections are the canonical projections onto the coordinates. Often it is denoted by $\ell^\infty((E_i))$. The coproduct of $\{E_i \mid i \in I\}$ is the linear subspace of $\mathsf{X}_{i \in I} E_i$ consisting of those $(x_i)_{i \in I}$ satisfying $\sum_{i \in I} \|x_i\| < \infty$ together with the componentwise injections. The norm of $(x_i)_{i \in I}$ is then defined by $\|(x_i)\| = \sum_{i \in I} \|x_i\|$, and for that reason the coproduct is typically denoted by $\ell^1((E_i))$.

Two special situations for product and coproduct need to be singled out. Suppose $I = \emptyset$. The "empty" product, if it exists, is the unique object $A \in \mathrm{ob}(\mathscr{C})$ with the property that for each $C \in \mathrm{ob}(\mathscr{C})$ there is a unique morphism $C \to A$; that is, $\mathrm{Mor}_\mathscr{C}(C, A)$ has precisely one element. In this case, A is called a *final object*. The "empty" coproduct is the unique object $A \in \mathrm{ob}(\mathscr{C})$, if it exists, satisfying: for each $C \in \mathrm{ob}(\mathscr{C})$ there is a unique morphism $A \to C$; that is, $\mathrm{Mor}_\mathscr{C}(A, C)$ has precisely one element. In this case, A is called an *initial object*.

For instance in $\mathscr{S}$, $\emptyset$ is the initial object and $\{\emptyset\}$ is the final object. In $R\text{-}\mathscr{M}od$, the zero module 0 serves as both the initial and the final object. In the category of fields, however, there are no initial or final objects (because a field homomorphism preserves the characteristic of a field).

An object in a category that is both initial and final is called a *zero object* As such an object is unique (cf. Exercise 2.7.12) it is often denoted as 0.

Some categories such as $R\text{-}\mathscr{M}od$ or $\mathscr{B}an_\infty$ have an additional structure on their morphism sets: they are abelian groups. In generalisation of this, one calls a category $\mathscr{C}$ *additive* if it has zero, finite coproducts and all morphism sets $\mathrm{Mor}(A, B)$, $A, B \in \mathrm{ob}(\mathscr{C})$ carry the structure of an abelian group such that the composition of morphisms is bilinear. In this case, finite products exist as well and agree with the finite coproducts.

2.7 Exercises

Exercise 2.7.1 Let R be a unital ring with centre $Z(R)$. Show that the mapping $a \mapsto L_a$, where, for $a \in R$, $L_a \colon R \to R$ is given by $L_a(x) = ax$, $x \in R$ provides isomorphisms between the rings R and $\mathrm{End}_R(R_R)$, respectively, and between $Z(R)$ and $\mathrm{End}_R({}_R R_R)$, respectively.

Exercise 2.7.2 Let R be a ring, and let ${}_R M \in R\text{-}\mathscr{M}od$. Show that the annihilator of ${}_R M$ in R,

$$\mathrm{Ann}_R({}_R M) = \{r \in R \mid r \cdot m = 0 \text{ for all } m \in {}_R M\}$$

is an ideal in R. Consider M as a left module over the quotient ring $S = R/\mathrm{Ann}_R(_RM)$ in the canonical way. Show that $_SM$ is a faithful left S-module.

Exercise 2.7.3 Let R be a ring and $_RM \in R\text{-}\mathcal{M}od$. If J is an ideal of R contained in $\mathrm{Ann}_R(_RM)$ then M carries a natural structure as a left R/J-module. Show that a subgroup of M is an submodule of $_RM$ if and only if it is a submodule of $_{R/J}M$.

Exercise 2.7.4 Let R be a ring, and let $_RN$, $_RM \in R\text{-}\mathcal{M}od$ with $_RN \leq _RM$. Show that, if both $_RN$ and $_RM/_RN$ are finitely generated, then $_RM$ is finitely generated.

Exercise 2.7.5 Show that the direct product and the direct sum of a family of left R-modules are uniquely determined up to isomorphism by the universal properties stated in Proposition 2.4.1.

Exercise 2.7.6 Let R be a unital ring, and let $_RM \in R\text{-}\mathcal{M}od$. Show that if $_RM$ is finitely generated and free then it is isomorphic to $_RR^n$ for some $n \in \mathbb{N}$.

Exercise 2.7.7 Let R be a unital ring, and let $_RM$, $_RN \in R\text{-}\mathcal{M}od$. Suppose that $_RN$ is free and let $f: {}_RM \to {}_RN$ be an R-epimorphism. Show that there exists $g \in \mathrm{Hom}_R({}_RN, {}_RM)$ such that $f \circ g = \mathrm{id}_N$ and that $_RM = \ker(f) \oplus \mathrm{im}(g)$ (that is, $_RM = \ker(f) + \mathrm{im}(g)$ and $\ker(f) \cap \mathrm{im}(g) = 0$).

Exercise 2.7.8 Prove the *modular law of module theory*:

$$_RL \cap (_RK + _RN) = {}_RK + (_RL \cap _RN)$$

whenever $_RK$, $_RL$, $_RN$ are submodules of $_RM \in R\text{-}\mathcal{M}od$ with $_RK \subseteq _RL$ for some ring R.

Exercise 2.7.9 Let $_RM \in R\text{-}\mathcal{M}od$ for a unital ring R. Show the following two statements.

(i) For each $m \in {}_RM$, we have $Rm \cong R/\mathrm{Ann}_R(m)$.
(ii) $\mathrm{Ann}_R(_RM) = \bigcap_{f \in \mathrm{Hom}_R(R,M)} \ker f$.

Exercise 2.7.10 Prove the identities in Eq. (2.6.1).

Exercise 2.7.11 Show that, if they exist, (arbitrary, non-empty) coproducts in a category are unique up to isomorphism.

Exercise 2.7.12 Let $\mathscr{C}$ be a category. Show that:

(i) All initial objects in $\mathscr{C}$ are isomorphic.
(ii) All final objects in $\mathscr{C}$ are isomorphic.
(iii) Hence a zero object is unique up to isomorphism, and we speak of *the* zero object.

The Isomorphism Theorems

3

Just as in group theory and in ring theory, we need to be able to manipulate modules freely under isomorphisms; this is done via the so-called isomorphism theorems which are discussed in this short chapter. The second part of this chapter deals with connections between categories, the functors, and with connections between the functors, the natural transformations.

3.1 Isomorphisms Between Modules

The basis for all the canonical isomorphism theorems is the following result.

3.1.1 Canonical Factorisation of Module Homomorphisms *Let $_RM$, $_RN \in$ R-Mod for some ring R. For every $f \subset \mathrm{Hom}_R(_RM, _RN)$ there exist a monomorphism $\iota \in \mathrm{Hom}_R(\mathrm{im}(f), _RN)$, an epimorphism $\pi \in \mathrm{Hom}_R(_RM, _RM/\ker(f))$ and an isomorphism $\hat{f} \colon _RM/\ker(f) \to \mathrm{im}(f)$ such that $f = \iota \circ \hat{f} \circ \pi$; that is, the following diagram is commutative*

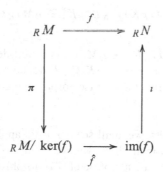

Proof We define the R-module homomorphism $\hat{f}$ by $\hat{f}(m + \ker(f)) = f(m)$, $m \in {}_R M$. Suppose $m + \ker(f) = 0$, that is, $m \in \ker(f)$. Then $f(m) = 0$ and therefore $\hat{f}$ is well defined. Suppose $\hat{f}(m + \ker(f)) = 0$; then $m \in \ker(f)$ and thus $m + \ker(f) = 0$. Therefore $\hat{f}$ is injective. By construction, $\hat{f}$ makes the above diagram commutative, where $\pi \colon {}_R M \to {}_R M / \ker(f)$ is the canonical epimorphism and $\iota \colon \mathrm{im}(f) \to {}_R N$ is the canonical monomorphism, and it is surjective onto $\mathrm{im}(f)$. As a result, it is the desired isomorphism. $\square$

We can now draw several nice consequences from the above theorem.

3.1.2 First Isomorphism Theorem *Let* ${}_R M$, ${}_R N \in R\text{-}\mathcal{M}od$ *for some ring R. For every $f \in \mathrm{Hom}_R ({}_R M, {}_R N)$, we have* ${}_R M / \ker(f) \cong \mathrm{im}(f)$.

Proof The mapping $\hat{f}$ in the canonical factorisation supplies the desired isomorphism. $\square$

3.1.3 Second Isomorphism Theorem *Let* ${}_R M$, ${}_R N$, ${}_R K \in R\text{-}\mathcal{M}od$ *for some ring R. Suppose* ${}_R N \leq {}_R M$ *and* ${}_R K \leq {}_R M$. *Then*

$$({}_R N + {}_R K)/{}_R K \cong {}_R N/({}_R N \cap {}_R K).$$

Proof Let $f \colon {}_R N \to ({}_R N + {}_R K)/{}_R K$ be the composition of the inclusion ${}_R N \to {}_R N + {}_R K$ and the canonical epimorphism from ${}_R N + {}_R K$ onto $({}_R N + {}_R K)/{}_R K$. For $n \in {}_R N$ and $k \in {}_R K$ we have $n + {}_R K = n + k + {}_R K$ and hence, f is surjective. Its kernel is $\ker(f) = \{n \in {}_R N \mid n \in {}_R K\} = {}_R N \cap {}_R K$. The first isomorphism theorem thus yields the result. $\square$

3.1.4 Third Isomorphism Theorem *Let* ${}_R M$, ${}_R K$, ${}_R L \in R\text{-}\mathcal{M}od$ *for some ring R. Suppose* ${}_R K \leq {}_R L \leq {}_R M$. *Then*

$$\left({}_R M/{}_R K\right)/\left({}_R L/{}_R K\right) \cong {}_R M/{}_R L.$$

Proof The mapping $f \colon {}_R M/{}_R K \to {}_R M/{}_R L$ is a well-defined epimorphism since $m + {}_R K = 0$ implies $m + {}_R L = 0$, as ${}_R K \subseteq {}_R L$. Its kernel is $\ker(f) = \{m + {}_R K \mid m \in {}_R L\} = {}_R L/{}_R K$; therefore, the first isomorphism theorem yields the statement.
$\square$

In the subsequent chapters, we shall see manifold applications of these isomorphism theorems. For the moment, however, we want to address a similar question for categories: Is there a convenient notion of "isomorphism" of categories? For this question to make sense, we first have to introduce connections between two different categories; these are the functors of the next section.

3.2   Functors and Natural Transformations

In this section we will discuss how 'to move' from one category to another; this is done via 'functions between categories', the functors. We shall then see how to move between functors; this is done via natural transformations.

Definition 3.2.1 Let $\mathscr{C}$ and $\mathscr{D}$ be categories. A *functor* $\mathsf{F} \colon \mathscr{C} \to \mathscr{D}$ consists of

(i) a mapping $A \mapsto \mathsf{F}(A)$ between the object classes of $\mathscr{C}$ and $\mathscr{D}$;
(ii) a family of mappings $f \mapsto \mathsf{F}(f)$ between the sets of morphisms

$$\mathrm{Mor}_{\mathscr{C}}(A, B) \to \mathrm{Mor}_{\mathscr{D}}(\mathsf{F}(A), \mathsf{F}(B)), \quad A, B \in \mathrm{ob}(\mathscr{C}) \ \textit{(covariant functor)}, \textit{or}$$

$$\mathrm{Mor}_{\mathscr{C}}(A, B) \to \mathrm{Mor}_{\mathscr{D}}(\mathsf{F}(B), \mathsf{F}(A)), \quad A, B \in \mathrm{ob}(\mathscr{C}) \ \textit{(contravariant functor)}$$

such that
(a) $\forall A \in \mathrm{ob}(\mathscr{C}) \colon \quad \mathsf{F}(1_A) = 1_{\mathsf{F}(A)}$;
(b) $\forall A, B, C \in \mathrm{ob}(\mathscr{C}) \ \forall f \in \mathrm{Mor}_{\mathscr{C}}(A, B), \ g \in \mathrm{Mor}_{\mathscr{C}}(B, C)$:

$$\mathsf{F}(gf) = \mathsf{F}(g)\mathsf{F}(f) \quad \text{(covariant) or } \mathsf{F}(gf) = \mathsf{F}(f)\mathsf{F}(g) \quad \text{(contravariant)}.$$

The functor F is called *faithful* if it is injective on morphism sets and it is called *full* if it is surjective on morphism sets.

Let us have a look at a number of examples of functors.

Examples 3.2.2

1. The *constant functor* $\mathscr{C} \to \mathscr{C}$, $A \mapsto A_0$ where A_0 is a fixed object in a category $\mathscr{C}$ sends each morphism f to the identity 1_{A_0} of A_0. It is both co- and contravariant.
2. The *identical functor* $\mathscr{C} \to \mathscr{C}$, $A \mapsto A$ and $f \mapsto f$ is an important covariant functor on every category.
3. The *forgetful functor*: A *concrete category* is a pair $(\mathscr{C}, \mathsf{F})$ consisting of a category $\mathscr{C}$ and a faithful functor $\mathsf{F} \colon \mathscr{C} \to \mathscr{S}$. Such a functor is an example of a *forgetful functor* as it 'forgets' some of the structure the objects in $\mathscr{C}$ may have by replacing them with the 'underlying' set. The morphisms in $\mathscr{C}$ correspond, via F, to set mappings on the 'underlying' sets. Such a functor is clearly covariant. Examples of concrete categories are $R\text{-}\mathscr{M}\!od$ or $\mathscr{T}\!op$.

   Similar forgetful functors can be defined from $\mathscr{C}\!omp$ to $\mathscr{T}\!op$ or $\mathscr{R}ing$ to $\mathscr{A}\mathscr{G}r$; in each case, part "of the structure on the objects" is forgotten and the corresponding requirement on the morphisms ignored.

4. The *inclusion functor*: Let $\mathscr{C}_0$ be a subcategory of the category $\mathscr{C}$. Then $\mathrm{ob}(\mathscr{C}_0) \ni$ $A \mapsto A \in \mathrm{ob}(\mathscr{C})$, $\mathrm{mor}(\mathscr{C}_0) \ni f \mapsto f \in \mathrm{mor}(\mathscr{C})$ is a covariant functor, the inclusion of the subcategory into the larger category.

5. The functor C: For a compact Hausdorff space X, let $C(X)$ denote the set of all continuous complex-valued functions on X. This set has a natural structure of a commutative unital ring using the pointwise operations. For φ, $\psi \colon X \to \mathbb{C}$ we define $\varphi + \psi$ and $\varphi\psi$ via $(\varphi + \psi)(x) = \varphi(x) + \psi(x)$ and $(\varphi\psi)(x) = \varphi(x)\psi(x)$, $x \in X$. The identity is the constant function $x \mapsto 1$. In this way, every object $X \in \mathrm{ob}(\mathscr{C}omp)$ is mapped onto an object $C(X) \in \mathrm{ob}(\mathscr{R}ing_1)$. Let $f \in \mathrm{Mor}(X, Y)$ for $X, Y \in \mathrm{ob}(\mathscr{C}omp)$. Then $C(f) \colon \varphi \mapsto \varphi \circ f$ defines a unital ring homomorphism from $C(Y)$ to $C(X)$. It is easy to check that this results in a contravariant functor C.

Since $C(X)$ in addition has the structure of a complex Banach space—the norm of φ is defined as $\|\varphi\| = \sup\{|\varphi(x)| \mid x \in X\}$—, there is a similar contravariant functor $C \colon \mathscr{C}omp \to \mathscr{B}an_1$.

We now come to a class of functors that are of paramount importance for us.

6. The Hom-*functors*: Let R be a unital ring, and let $\mathscr{C}$ be a full subcategory of R-$\mathscr{M}od$. Fix $_RM \in R$-$\mathscr{M}od$. Then

$$\mathrm{Hom}_R\,(_RM, -) \colon \mathscr{C} \longrightarrow \mathscr{A}\mathscr{G}r$$

$$_RN \longmapsto \mathrm{Hom}_R\,(_RM, _RN) \qquad\qquad (3.2.1)$$

$$\mathrm{Hom}_R\,(_RN, _RL) \ni f \longmapsto f_* = \mathrm{Hom}_R\,(_RM, f)$$

given by $f_*(g) = fg$ for $g \in \mathrm{Hom}_R\,(_RM, _RN)$ yields the covariant Hom-functor.

Note that $f_* \colon \mathrm{Hom}_R\,(_RM, _RN) \to \mathrm{Hom}_R\,(_RM, _RL)$ is a homomorphism of abelian groups.

Similarly to (3.2.1) we get the contravariant Hom-functor, for fixed $_RM \in R$-$\mathscr{M}od$,

$$\mathrm{Hom}_R\,(-, _RM) \colon \mathscr{C} \longrightarrow \mathscr{A}\mathscr{G}r$$

$$_RN \longmapsto \mathrm{Hom}_R\,(_RN, _RM) \qquad\qquad (3.2.2)$$

$$\mathrm{Hom}_R\,(_RN, _RL) \ni f \longmapsto f^* = \mathrm{Hom}_R\,(f, _RM)$$

given by $f^*(g) = gf$ for $g \in \mathrm{Hom}_R\,(_RL, _RM)$.

As above, $f^*\colon \operatorname{Hom}_R({}_RL, {}_RM) \to \operatorname{Hom}_R({}_RN, {}_RM)$ is a group homomorphism.

It is plain that the above concepts can be generalised to other categories; cf., e.g., the functor C above. A functor $\mathsf{F}\colon \mathscr{C} \to \mathscr{S}$ is called *representable* if there is an object $A \in \operatorname{ob}(\mathscr{C})$ such that $\mathsf{F} = \operatorname{Mor}_{\mathscr{C}}(A, -)$ in the covariant case and $\mathsf{F} = \operatorname{Mor}_{\mathscr{C}}(-, A)$ in the contravariant case. In this situation, A is called a *representing object*.

We now turn our attention to the 'functions between functors'.

Let F and G be functors of the same variance between the categories $\mathscr{C}$ and $\mathscr{D}$. For each $A \in \operatorname{ob}(\mathscr{C})$, let $\eta_A \in \operatorname{Mor}_{\mathscr{D}}(\mathsf{F}(A), \mathsf{G}(A))$. Then $\eta = (\eta_A)$ is a *natural transformation from* F *to* G if, for each $A, B \in \operatorname{ob}(\mathscr{C})$ and each $f \in \operatorname{Mor}_{\mathscr{C}}(A, B)$, the following diagram is commutative

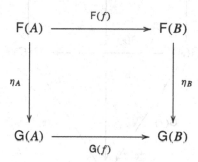

covariant case

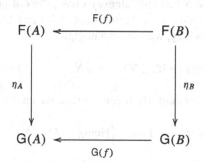

contravariant case

In the case that every η_A is an isomorphism, η is called a *natural isomorphism*.

A natural transformation shifts commuting triangles from one category to the other; the diagram below illustrates the covariant case (where $f \in \operatorname{Mor}_{\mathscr{C}}(A, B)$, $g \in \operatorname{Mor}_{\mathscr{C}}(B, C)$).

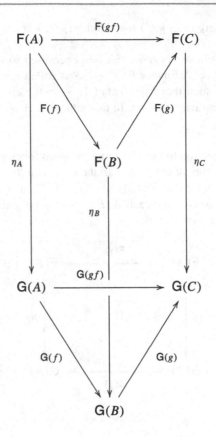

Example 3.2.3 Let $\mathscr{C}$ be a full subcategory of $R\text{-}\mathscr{Mod}$. Let $_RM, _RN \in \mathscr{C}$. For fixed $m \in {}_RM$, a morphism $f \in \mathrm{Hom}_R\,({}_RM, {}_RN)$ can be evaluated at m; this gives $f(m) \in {}_RN$. In this way, we obtain a mapping

$$_RM \times \mathrm{Hom}_R\,({}_RM, {}_RN) \longrightarrow {}_RN, \quad (m, f) \mapsto f(m);$$

this is called the *evaluation map*. By means of this, we have a mapping

$$\eta_M : {}_RM \longrightarrow \mathrm{Hom}_{\mathbb{Z}}\left(\mathrm{Hom}_R\,({}_RM, {}_RN), {}_RN\right)$$

as each map $f \mapsto f(m)$ is a group homomorphism from $\mathrm{Hom}_R\,({}_RM, {}_RN)$ into $_RN$.

The Hom-functors $\mathrm{Hom}_R\,(-, {}_RN)$ and $\mathrm{Hom}_{\mathbb{Z}}\,(-, {}_RN)$ are contravariant; thus their composition $\mathsf{G} = \mathrm{Hom}_{\mathbb{Z}}\left(\mathrm{Hom}_R\,(-, {}_RN), {}_RN\right)$ is a covariant functor from $\mathscr{C}$ into $\mathscr{AGr}$. Let $\mathsf{F} : \mathscr{C} \to \mathscr{AGr}$ be the forgetful functor. Then

$$\eta_M : \mathsf{F}({}_RM) \longrightarrow \mathsf{G}({}_RM) \qquad ({}_RM \in \mathscr{C})$$

defines a natural transformation between F and G. To verify this statement, we need to check the commutativity of the following diagram

$$
\begin{array}{ccc}
F(_RM) & \xrightarrow{\ \eta_M\ } & \mathrm{Hom}_{\mathbb{Z}}\left(\mathrm{Hom}_R\,(_RM,\,_RN),\,_RN\right) = G(_RM) \\[2mm]
F(g) \Big\downarrow & & \Big\downarrow G(g) \\[2mm]
F(_RL) & \xrightarrow{\ \eta_L\ } & \mathrm{Hom}_{\mathbb{Z}}\left(\mathrm{Hom}_R\,(_RL,\,_RN),\,_RN\right) = G(_RL)
\end{array}
$$

Let $m \in M$, $g \in \mathrm{Hom}_R\,(_RM,\,_RL)$, $f \in \mathrm{Hom}_R\,(_RL,\,_RN)$. Then

$$
\eta_L(F(g)(m))(f) = f(g(m)) = (fg)(m) = g^*(f)(m)
$$
$$
= \eta_M(m)(g^*(f)) = (\eta_M(m)g^*)(f) = G(g)(\eta_M(m))(f)
$$

which shows that the diagram is commutative.

Suppose $\mathscr{C}$ and $\mathscr{D}$ are categories and $F\colon \mathscr{C} \to \mathscr{D}$ and $G\colon \mathscr{D} \to \mathscr{C}$ are functors such that $F \circ G$ is naturally isomorphic to the identity functor on $\mathscr{D}$ and $G \circ F$ is naturally isomorphic to the identity functor on $\mathscr{C}$. Then the two categories are *naturally equivalent*; in this way we can compare categories of quite different 'sizes' with each other.

Example 3.2.4 A *commutative unital C*-algebra* A consists of a commutative unital ring A which at the same time is a complex Banach space with a norm $\|\cdot\|$. In addition, we require that the norm is submultiplicative, that is, for $x, y \in A$ we have $\|xy\| \le \|x\|\,\|y\|$, and that $\|1\| = 1$. Moreover, there exists an *involution* $*$ on A, that is, a conjugate-linear anti-multiplicative mapping $*\colon A \to A$ such that $(x^*)^* = x$ and $\|xx^*\| = \|x\|^2$ for all $x \in A$. It is easy to verify that, for every compact Hausdorff space X, the space $C(X)$ is a commutative unital C*-algebra, where the involution is given by $\psi^*(x) = \overline{\varphi(x)}$ for each $x \in X$; compare Example 3.2.2 above. We define a morphism ρ between two commutative unital C*-algebras A and B to be a linear mapping which is also a unital ring homomorphism and satisfies $\rho(x^*) = \rho(x)^*$ for all $x \in A$. It turns out that every such mapping is a bounded linear operator of norm 1 and hence the isomorphisms are linear isometries. The category built in this fashion will be denoted by $\mathscr{AC}_1^*$. It is a non-full subcategory of $\mathscr{B}an_1$. (For more information on C*-algebras, we refer the reader to [1] and [32].) As a result, the above functor C can also be defined from $\mathscr{Comp}$ into $\mathscr{AC}_1^*$.

Let $A \in \mathrm{ob}(\mathscr{AC}_1^*)$. Endowed with the hull-kernel topology the set of all maximal ideals of A becomes a compact Hausdorff space, see [1] and [32], which we shall denote by $\Delta(A)$. Let $\rho\colon A \to B$ be a morphism in $\mathscr{AC}_1^*$. For each maximal ideal $I \subseteq B$ the inverse image $\rho^{-1}(I)$ is a maximal ideal in A, we shall denote it by $\rho^*(I)$. It can be shown that the mapping $\rho^*\colon \Delta(B) \to \Delta(A)$ so defined is continuous, i.e., a morphism in $\mathcal{Comp}$. In addition, Δ becomes a contravariant functor from $\mathcal{Comp}$ into $\mathcal{AC}_1^*$.

Gelfand theory now establishes that $\Delta(C(X))$ is naturally homeomorphic to X, for every $X \in \mathrm{ob}(\mathcal{Comp})$, and that $C(\Delta(A))$ is naturally isomorphic to A, for every $A \in \mathrm{ob}(\mathcal{AC}_1^*)$. By means of this, we obtain a natural equivalence between the categories $\mathcal{Comp}$ and $\mathcal{AC}_1^*$.

Far more detailed information on functors and natural transformations can be found in [25] and [30], for example.

3.3 Exercises

We use this opportunity to introduce an important device from homological algebra. A sequence

$$\cdots \xrightarrow{f_{j-1}} {}_R M_j \xrightarrow{f_j} {}_R M_{j+1} \xrightarrow{f_{j+1}} {}_R M_{j+2} \xrightarrow{f_{j+2}} \cdots$$

$$(3.3.1)$$

consisting of a sequence $({}_R M_j)_{j \in \mathbb{N}}$ of modules in $R\text{-}\mathcal{Mod}$ and module maps $f_j \in \mathrm{Hom}_R({}_R M_j, {}_R M_{j+1})$, $j \in \mathbb{N}$ is called *exact at step* $j + 1$ if $\ker(f_{j+1}) = \mathrm{im}(f_j)$. The entire sequence is called an *exact sequence* if it is exact at every step $j \in \mathbb{N}$.

In case the sequence consists of three (non-trivial) modules and the zero module at either end,

$$0 \xrightarrow{\quad} {}_R M \xrightarrow{f} {}_R N \xrightarrow{g} {}_R L \xrightarrow{\quad} 0 \qquad (3.3.2)$$

and is exact we speak of a *short exact sequence*. Note that we do not have to specify the module maps at either end (they can only be the zero maps) and that exactness in this situation means that ${}_R M \xrightarrow{f} {}_R N$ is injective, ${}_R N \xrightarrow{g} {}_R L$ is surjective and $\ker(g) = \mathrm{im}(f)$. By the first isomorphism theorem, it follows that ${}_R L \cong {}_R N / \mathrm{im}(f)$ so one can think of ${}_R L$ as a quotient of ${}_R N$. The degenerate case when ${}_R L = 0$ means that f itself is an isomorphism.

Exercise 3.3.1 Let ${}_R N, {}_R L$ be submodules of the module ${}_R M \in R\text{-}\mathcal{Mod}$. Define canonical module mappings to obtain a short exact sequence

$$0 \xrightarrow{\quad} {}_R M/({}_R N \cap {}_R L) \xrightarrow{\quad} {}_R M/{}_R N \times {}_R M/{}_R L \xrightarrow{\quad} {}_R M/({}_R N + {}_R L) \xrightarrow{\quad} 0.$$

Then apply this short exact sequence to obtain an isomorphism

$$({}_R N + {}_R L)/({}_R N \cap {}_R L) \cong ({}_R N + {}_R L)/{}_R N \times ({}_R N + {}_R L)/{}_R L.$$

Exercise 3.3.2 Show that $\text{Hom}_R\,(_R M, -)$ for fixed $_R M \in R\text{-}\mathcal{M}od$ is a covariant functor from $R\text{-}\mathcal{M}od$ into the category of abelian groups.

Exercise 3.3.3 Show that a morphism f in $R\text{-}\mathcal{M}od$ is a monomorphism if and only if $f_* = \text{Hom}_R\,(_R M, f)$ is injective for every $_R M \in R\text{-}\mathcal{M}od$ and f is an epimorphism if and only if $f^* = \text{Hom}_R\,(f, _R M)$ is injective for every $_R M \in R\text{-}\mathcal{M}od$.

Exercise 3.3.4 In Chap. 2, p. 23 we introduced the notion of an additive category. A functor F between two additive categories $\mathscr{C}$ and $\mathscr{D}$ is called *additive* if $\text{F}(f + g) = \text{F}(f) + \text{F}(g)$ for any two morphisms f, g in the same morphism set in $\mathscr{C}$. Show that the Hom-functor $\text{Hom}_R\,(_R M, -)$ for fixed $_R M \in R\text{-}\mathcal{M}od$ is additive from $R\text{-}\mathcal{M}od$ into $\mathscr{A}\mathscr{G}r$.

Exercise 3.3.5 In Exercise 1.3.3 we saw how to turn an S-module into an R-module given a fixed ring homomorphism $\rho\colon R \to S$ between two rings R and S. Does this procedure give us a covariant functor $\text{F}_\rho\colon S\text{-}\mathcal{M}od \to R\text{-}\mathcal{M}od$?

Exercise 3.3.6 Let R be a unital ring and let $_R L$, $_R N$ be submodules of the right R-module $_R M$. We define $[_R L : {}_R N] = \{a \in R \mid a \cdot {}_R N \subseteq {}_R L\}$. Show the following statements:

(i) $\lfloor_R L : {}_R N\rfloor = R$ if $_R N \subseteq {}_R L$.
(ii) $[_R L_1 \cap {}_R L_2 : {}_R N] = [_R L_1 : {}_R N] \cap [_R L_2 : {}_R N]$ for any $_R L_1, {}_R L_2 \le {}_R M$.
(iii) $[_R L : {}_R L + {}_R N] = [_R L : {}_R N]$.

Exercise 3.3.7 Let R be a unital commutative ring and let $_R M \in R\text{-}\mathcal{M}od$. Using the same notation as in the previous exercise, show that the following two conditions are equivalent:

(a) For all $0 \ne {}_R N \le {}_R M$, $[0 : {}_R N] = [0 : {}_R M]$;
(b) For all ideals $I \trianglelefteq R$ and $0 \ne {}_R N \le {}_R M$, $I \cdot {}_R N = 0$ implies that $I \cdot {}_R M = 0$.

Show further that, if I is an ideal of R, then the quotient module $_R R/_R I$ has any of the above two propeties if and only if I is a prime ideal.

Noetherian Modules

4

Finiteness conditions on the ideal structure of a ring allow us to develop a more detailed structure theory. In the noncommutative framework, it is more natural to do this for (one-sided) modules rather than for (one-sided) ideals, so we shall follow this approach right from the beginning.

The following result is fundamental to this chapter.

Theorem 4.1 *Let $_RM \in R\text{-}\mathscr{M}od$ for a unital ring R. The following conditions are equivalent.*

(a) $_RM$ *satisfies ACC, the ascending chain condition:*
   Each ascending chain $_RN_1 \subseteq {}_RN_2 \subseteq \ldots \subseteq {}_RN_i \subseteq \ldots$ of submodules $_RN_i \leq$
   $_RM$ *becomes stationary, i.e., there exists k such that $_RN_i = {}_RN_k$ for all $i \geq k$.*
(b) *Each strictly ascending chain $_RN_1 \subset {}_RN_2 \subset \ldots \subset {}_RN_i \subset \ldots$ of submodules*
   $_RN_i \leq {}_RM$ *is finite.*
(c) *Every non-empty set of submodules of $_RM$ contains a maximal element.*
(d) *Every submodule of $_RM$ is finitely generated.*

Proof The implication (a) $\Rightarrow$ (b) is trivial. Towards (b) $\Rightarrow$ (c) let $\mathscr{N}$ be a non-empty set of submodules of $_RM$ without a maximal element. Take $_RN_1 \in \mathscr{N}$; then there is $_RN_2 \in \mathscr{N}$ with $_RN_1 \subset {}_RN_2$. Let $_RN_1, \ldots, {}_RN_k \in \mathscr{N}$ be chosen such that $_RN_1 \subset {}_RN_2 \subset \ldots \subset {}_RN_k$. Since $_RN_k$ is not maximal in $\mathscr{N}$, there is $_RN_{k+1} \in \mathscr{N}$ strictly containing $_RN_k$. By induction, we obtain an infinite strictly ascending chain of submodules of $_RM$. Thus (b) fails.

(c) $\Rightarrow$ (d) Let $_RN \leq {}_RM$. Let $\mathscr{N}$ be the set of all finitely generated submodules of $_RN$. As these are submodules of $_RM$, $\mathscr{N}$ has a maximal element $_RN_0$. If there is $n \in {}_RN \setminus {}_RN_0$ then $_RN_0 + Rn$ is a finitely generated submodule of $_RN$ strictly larger than $_RN_0$. As this contradicts the maximality of $_RN_0$ in $\mathscr{N}$, we find that $_RN = {}_RN_0$ is finitely generated.

M. Mathieu, *Classically Semisimple Rings*,
https://doi.org/10.1007/978-3-031-14209-3_4

(d) $\Rightarrow$ (a) Let $_RN_1 \subseteq {_R}N_2 \subseteq \ldots \subseteq {_R}N_i \subseteq \ldots$ be an ascending chain of submodules of $_RM$. Set $_RN = \bigcup_i {_R}N_i$ which is a submodule of $_RM$. By hypothesis, $_RN$ is finitely generated. Let $\{m_1, \ldots, m_k\}$ be a set of generators of $_RN$. Then there is $j \in \mathbb{N}$ such that $\{m_1, \ldots, m_k\} \subseteq {_R}N_j$. It follows that $_RN = {_R}N_j$, equivalently, $_RN_i = {_R}N_j$ for all $i \geq j$. $\square$

Definition 4.2 A left R-module $_RM \in R\text{-}\mathcal{M}\!od$ for a unital ring R is called *Noetherian* if it satisfies any, and hence every, condition in Theorem 4.1 above. The ring R is called *left Noetherian* if the module $_RR$ is Noetherian.

Historical Note Born in 1882 in Erlangen (Bavaria, Germany), Emmy Amalie Noether was one of the first female students at a German university. After she had become a certificated teacher of English and French in Bavarian girls schools, she studied Mathematics in Erlangen and Göttingen where she attended lectures by David Hilbert, Felix Klein and Hermann Minkowski, amongst others. In 1907 she received her doctorate under Paul Gordon. Many bureaucratic problems had to be overcome before Emmy Noether achieved the position of Privatdozent at the University in Göttingen in 1919. She worked on invariant theory and mathematical physics and became instrumental in the development of abstract algebra. Her influence on ring theory and especially finiteness conditions on ideals cannot be overestimated. Twice she gave invited addresses to the International Congress of Mathematicians, in Bologna in 1928 and in Zürich in 1932. She died in 1935 in Bryn Mawr (Pennsylvania, USA) where she had to emigrate to because of her Jewish ancestry.

A detailed account of Emmy Noether's life and work, written by J. J. O'Connor and E. F. Robinson can be found at

https://mathshistory.st-andrews.ac.uk/Biographies/Noether_Emmy/

There is, of course, an analogous concept of Noetherian right modules and of a right Noetherian ring. These two concepts, however, differ from each other as the following example illustrates.

Example 4.3 Let s and t be two symbols and let $R = \mathbb{Z}\langle s, t \mid t^2 = ts = 0\rangle$ be the quotient of the free unital ring $\mathbb{Z}\langle s, t\rangle$ by the stated relations. As every element of R is a unique combination of words of the form s^n and $s^n t$, $n \geq 0$, R can be written as $R = \mathbb{Z}[s] \oplus \mathbb{Z}[s]t$. The commutative ring $\mathbb{Z}[s]$ is Noetherian by Hilbert's Basis Theorem (see below). The ideal $\mathbb{Z}[s]t$ of R is considered as a $\mathbb{Z}[s]$-module; therefore the finitely generated $\mathbb{Z}[s]$-module $\mathbb{Z}[s] \oplus \mathbb{Z}[s]t$ is Noetherian, by Corollary 4.1.4 below, in other words, R is left Noetherian.

In order to show that R is not right Noetherian, it suffices to verify that the submodule $\mathbb{Z}[s]t \in \mathcal{M}\!od\text{-}R$ is not finitely generated. The monomials $s^n t$, $n \geq 0$ are $\mathbb{Z}$-linearly independent; thus there cannot exist finitely many $p_1, \ldots, p_k \in \mathbb{Z}[s]t$ such that an arbitrary element $p = \sum_{n \geq 0} m_n s^n t$, where $m_n \in \mathbb{Z}$ are non-zero for at most finitely many n, can be written as $p = \sum_{j=1}^{k} m_j p_j$ as the latter would imply that $\mathbb{Z}[s]t$ is finitely generated as an abelian group. As a result, R is not right Noetherian.

A unital ring is said to be *Noetherian* if it is both left and right Noetherian. Evidently, every principal ideal domain is Noetherian. Far more general examples are provided by the following fundamental result.

4.4 Hilbert's Basis Theorem *For every left Noetherian ring R, the polynomial ring $R[x_1, \ldots, x_n]$ is left Noetherian.*

Proof Since $R[x, y] = R[x][y]$ it suffices to establish the result for polynomial rings in one indeterminate and then proceed by induction. Suppose that $R[x]$ is not left Noetherian. Take a left ideal I in $R[x]$ which is not finitely generated. Let $f_1 \in I$ be a polynomial with minimal degree. Then $I_1 = R[x]f_1 \subsetneq I$. Take $f_2 \in I \setminus I_1$ of minimal degree. Put $I_2 = R[x]f_1 + R[x]f_2 \subsetneq I$. Since I is not finitely generated, by induction we obtain an infinite strictly ascending chain $(I_n)_{n \in \mathbb{N}}$ of left ideals in $R[x]$. Let $n_k = \deg f_k$, $k \in \mathbb{N}$ and let a_{n_k} be the leading coefficient in f_k (i.e., a_{n_k} is the coefficient of x^{n_k}). Observe that $n_1 \leq n_2 \leq \ldots \leq n_k \leq n_{k+1} \leq \ldots$. Let $L_k = Ra_{n_1} + \ldots + Ra_{n_k}$, $k \in \mathbb{N}$. Clearly, $(L_k)_{k \in \mathbb{N}}$ is an ascending chain of left ideals of R. We claim that this sequence is strictly ascending (hence R is not left Noetherian).

Suppose on the contrary that $a_{n_{k+1}} \in L_k = Ra_{n_1} + \ldots + Ra_{n_k}$ for some k. Hence, $a_{n_{k+1}} = \sum_{i=1}^{k} r_i a_{n_i}$ for some $r_i \in R$. Put

$$f(x) = f_{k+1}(x) - \sum_{i=1}^{k} r_i x^{n_{k+1} - n_i} f_i(x).$$

Then $f \in I_{k+1}$, the left ideal generated by $f_1, \ldots, f_k, f_{k+1}$ but $f \notin I_k$. We note that $\deg f < \deg f_{k+1}$:

$$a_{n_{k+1}} x^{n_{k+1}} - \sum_{i=1}^{n} r_i x^{n_{k+1} - n_i} a_{n_i} x^{n_i} = \left(a_{n_{k+1}} - \sum_{i=1}^{n} r_i a_{n_i} \right) x^{n_{k+1}} = 0.$$

This violates the choice of f_{k+1} being a polynomial in $I_{k+1} \setminus I_k$ of lowest degree. As a result our initial assumption entails that R is not left Noetherian. $\square$

4.1 Permanence Properties of Noetherian Modules

In this section, we shall study how Noetherian modules behave under various canonical constructions, such as taking submodules or homomorphic images. The key result is the following theorem in which we employ the concept of an exact sequence of modules, see p. 32.

Theorem 4.1.1 *Let R be a unital ring. Let*

$$0 \longrightarrow {}_R M_1 \xrightarrow{\;\;f\;\;} {}_R M_2 \xrightarrow{\;\;g\;\;} {}_R M_3 \longrightarrow 0$$

be a short exact sequence in R-$\mathcal{M}od$. Then ${}_R M_2$ is Noetherian if and only if both ${}_R M_1$ and ${}_R M_3$ are Noetherian.

Proof Suppose at first that ${}_R M_2$ is Noetherian. Every ascending chain of sub-modules of ${}_R M_1$ becomes an ascending chain of submodules of ${}_R M_2$ via the injective module map f. Hence it becomes stationary, and thus ${}_R M_1$ is Noetherian. Every ascending chain of submodules of ${}_R M_3$ is the image of an ascending chain of submodules of ${}_R M_2$ under the surjective module map g. Hence it becomes stationary, and thus ${}_R M_3$ is Noetherian.

Suppose now that both ${}_R M_1$ and ${}_R M_3$ are Noetherian, and let $\left({}_R N^{(k)}\right)_{k \in \mathbb{N}}$ be an ascending chain of submodules of ${}_R M_2$. Setting ${}_R N_1^{(k)} = f^{-1}({}_R N^{(k)})$ and ${}_R N_3^{(k)} = g({}_R N^{(k)})$, $k \in \mathbb{N}$ we obtain ascending chains of submodules of ${}_R M_1$ and of ${}_R M_3$, respectively. By hypothesis, there is $m \in \mathbb{N}$ such that ${}_R N_1^{(k)} = {}_R N_1^{(m)}$ and ${}_R N_3^{(k)} = {}_R N_3^{(m)}$ for all $k \geq m$. We want to show that ${}_R N^{(k)} = {}_R N^{(m)}$ for all $k \geq m$ and employ the commutative diagram below to this end.

$$
\begin{array}{ccccccccc}
0 & \longrightarrow & {}_R M_1 & \xrightarrow{\;f\;} & {}_R M_2 & \xrightarrow{\;g\;} & {}_R M_3 & \longrightarrow & 0 \\
& & \uparrow & & \uparrow & & \uparrow & & \\
0 & \longrightarrow & {}_R N_1^{(k)} & \xrightarrow{f^{(k)}} & {}_R N^{(k)} & \xrightarrow{g^{(k)}} & {}_R N_3^{(k)} & \longrightarrow & 0 \\
& & \| & & \uparrow & & \| & & \\
0 & \longrightarrow & {}_R N_1^{(m)} & \xrightarrow{f^{(m)}} & {}_R N^{(m)} & \xrightarrow{g^{(m)}} & {}_R N_3^{(m)} & \longrightarrow & 0
\end{array}
$$

where $m \leq k$ and $f^{(k)}$ and $g^{(k)}$ are the restrictions of f and g, respectively to the corresponding submodules. By construction, all rows are short exact sequences. As ${}_R N^{(m)} \subseteq {}_R N^{(k)}$ it remains to prove the reverse inclusion.

Let $x \in {}_R N^{(k)}$. Since ${}_R N_3^{(k)} = {}_R N_3^{(m)}$ there is $y \in {}_R N^{(m)}$ such that $g^{(k)}(x) = g^{(k)}(y)$. Since $\ker(g^{(k)}) = \operatorname{im}(f^{(k)})$ there is $y' \in {}_R N_1^{(k)}$ such that $x - y = f^{(k)}(y')$. As ${}_R N_1^{(k)} = {}_R N_1^{(m)}$, $y' \in {}_R N_1^{(m)}$. It follows that

$$f(y') = f^{(m)}(y') \in f^{(m)}({}_R N_1^{(m)}) = f(f^{-1}({}_R N^{(m)})) \subseteq {}_R N^{(m)}$$

and hence $x = y + f(y') \in {}_R N^{(m)}$ as required. $\qquad\square$

The above argument is an example of a 'diagram chase'.

Applying Theorem 4.1.1 to the short exact sequence

$$0 \longrightarrow {}_R N \longrightarrow {}_R M \longrightarrow {}_R M/{}_R N \longrightarrow 0,$$

where ${}_R N \leq {}_R M$ we immediately obtain the following result.

Corollary 4.1.2 *Let R be a unital ring. Let ${}_R N$ be a submodule of ${}_R M \in R\text{-}\mathcal{M}od$. Then ${}_R M$ is Noetherian if and only if both ${}_R N$ and ${}_R M/{}_R N$ are Noetherian.*

From the above theorem we obtain a wealth of further Noetherian modules.

Corollary 4.1.3 *Every finite direct sum of Noetherian modules is Noetherian.*

Proof Using induction, it suffices to consider the case of two Noetherian modules ${}_R L$ and ${}_R N$. In this case, the statement follows directly from Theorem 4.1.1 applied to the short exact sequence

$$0 \longrightarrow {}_R N \longrightarrow {}_R N \oplus {}_R L \longrightarrow {}_R L \longrightarrow 0.$$

□

Corollary 4.1.4 *Every finitely generated unital left module over a left Noetherian ring is Noetherian.*

Proof Let R be a left Noetherian ring. By Corollary 4.1.3, the module ${}_R R^n$ is Noetherian for every $n \in \mathbb{N}$. Let ${}_R M \subset R\text{-}mod$ and suppose $\{m_1, \ldots, m_n\}$ is a set of generators of ${}_R M$. The mapping

$$ {}_R R^n \longrightarrow {}_R M, \quad (r_1, \ldots, r_n) \longmapsto \sum_{i=1}^{n} r_i \cdot m_i $$

is an epimorphism; thus Corollary 4.1.2 yields the claim. □

We observe therefore that a unital ring R is left Noetherian if and only if every module in $R\text{-}mod$ is Noetherian.

4.2 Exact Categories and Exact Functors

The concept of an exact sequence is fundamental in the theory of modules. In this section we shall explore to what extent this tool can be employed in more general categories.

4.2.1 Kernels and Cokernels

Let $\mathscr{C}$ be a category with zero object, see Chap. 2, p. 23. We will denote this object (which is unique up to isomorphism by Exercise 2.7.12) by 0. Let $B, C \in \mathrm{ob}(\mathscr{C})$. The unique morphisms $f : 0 \to B$ and $g : C \to 0$ have the properties

(i) $h_1 f = h_2 f$ for all $h_1, h_2 \in \mathrm{Mor}_{\mathscr{C}}(B, A)$, $A \in \mathrm{ob}(\mathscr{C})$;
(ii) $g h_1 = g h_2$ for all $h_1, h_2 \in \mathrm{Mor}_{\mathscr{C}}(A, C)$, $A \in \mathrm{ob}(\mathscr{C})$;
(iii) $fg : C \to B$ is the unique morphism, denoted by $0_{(C,B)}$, satisfying

$$h\, 0_{(C,B)} = 0_{(C,A)} \quad \text{and} \quad 0_{(C,B)} k = 0_{(A,B)}$$

for all $A \in \mathrm{ob}(\mathscr{C})$ and $h \in \mathrm{Mor}_{\mathscr{C}}(B, A)$, $k \in \mathrm{Mor}_{\mathscr{C}}(A, C)$.

The above properties are easily checked (see Exercise 4.3.8 below) and the above-mentioned unique morphism is called a *zero morphism*. As a result, a category $\mathscr{C}$ with zero object has zero morphisms between all its objects.

In the sequel, we will assume that $\mathscr{C}$ has a zero object 0 and denote the zero morphisms unambiguously by 0, too.

Let $f \in \mathrm{Mor}_{\mathscr{C}}(A, B)$ for some $A, B \in \mathrm{ob}(\mathscr{C})$.

(i) A morphism $i : K \to A$ is a *kernel of* f if $fi = 0$ and for each $D \in \mathrm{ob}(\mathscr{C})$ and $g \in \mathrm{Mor}_{\mathscr{C}}(D, A)$ with $fg = 0$ there is a unique $h \in \mathrm{Mor}_{\mathscr{C}}(D, K)$ making the diagram below commutative

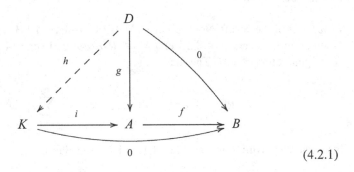

$$(4.2.1)$$

Any kernel is a monomorphism and is, up to isomorphism, unique.

(ii) A morphism $p\colon B \to C$ is a *cokernel of* f if $pf = 0$ and for each $D \in \mathrm{ob}(\mathscr{C})$ and $g \in \mathrm{Mor}_{\mathscr{C}}(B, D)$ with $gf = 0$ there is a unique $h \in \mathrm{Mor}_{\mathscr{C}}(C, D)$ making the diagram below commutative

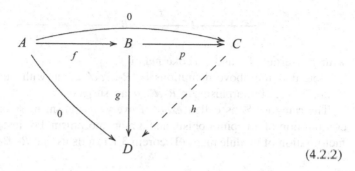

$$(4.2.2)$$

Any cokernel is an epimorphism and is, up to isomorphism, unique.

Since kernel and cokernel of a morphism f, if they exist, are unique, it suffices to give their domain and codomain, respectively, a symbol; these are, respectively, $\ker(f)$ and $\mathrm{cok}(f)$.

For instance, in $R\text{-}\mathcal{M}od$, the kernel of $f \in \mathrm{Hom}_R ({}_R M, {}_R N)$ is the usual kernel $\ker(f)$ together with the canonical embedding $\ker(f) \to {}_R M$, and the cokernel of f is the canonical projection $\pi\colon {}_R N \to {}_R N / \mathrm{im}(f)$. However, in $R\text{-}mod$, kernels need not exist since a submodule of a finitely generated module may be not finitely generated.

We are now in a position to introduce the concept of an exact sequence.

4.2.2 Exact Categories

Let $\mathscr{C}$ be a category with zero, kernels and cokernels.

A sequence $A \xrightarrow{f} B \xrightarrow{g} C$ in $\mathscr{C}$ is called *exact* if $gf = 0$ and in the commutative diagram below the unique morphism h is an epimorphism

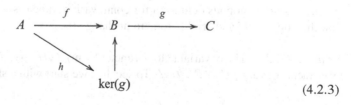

$$(4.2.3)$$

A sequence $\cdots \xrightarrow{f_i} A_i \xrightarrow{f_{i+1}} A_{i+1} \longrightarrow \cdots$ is *exact* if each 3-point part of it is exact. A *short exact sequence* in $\mathscr{C}$ is of the form

$$0 \longrightarrow A \xrightarrow{f} B \xrightarrow{g} C \longrightarrow 0$$

with f a kernel of g and g a cokernel of f.

Note that the above definitions in $R\text{-}\mathcal{M}od$ agree with our previous ones in Chap. 3 since epimorphisms in $R\text{-}\mathcal{M}od$ are surjective.

The category $\mathscr{C}$ is called *exact* if every morphism in $\mathscr{C}$ can be written as a composition of an epimorphism and a monomorphism. For instance, the canonical factorisation of module maps (Theorem 3.1.1) tells us that $R\text{-}\mathcal{M}od$ is exact.

4.2.3 Exact Functors

Let $\mathscr{C}$ and $\mathscr{D}$ be exact categories. A covariant functor $F\colon \mathscr{C} \to \mathscr{D}$ is said to be *left exact*, respectively *right exact*, if $F(0) = 0$ and for every short exact sequence

$$0 \longrightarrow A \xrightarrow{f} B \xrightarrow{g} C \longrightarrow 0$$

in $\mathscr{C}$ the sequence

$$0 \longrightarrow F(A) \xrightarrow{F(f)} F(B) \xrightarrow{F(g)} F(C)$$

is exact in $\mathscr{D}$, respectively the sequence

$$F(A) \xrightarrow{F(f)} F(B) \xrightarrow{F(g)} F(C) \longrightarrow 0$$

is exact in $\mathscr{D}$. If F is both left and right exact we say that F is *exact*.

There is an analogous definition for contravariant functors with arrows reversed (but the sides "left" and "right" remain unchanged!).

Example 4.2.3.1 The covariant Hom-functor $\mathrm{Hom}_R\,(_RU, -)\colon R\text{-}\mathcal{M}od \to \mathcal{AGr}$ is left exact for every $_RU \in R\text{-}\mathcal{M}od$. To see this, we start with a short exact sequence

$$0 \longrightarrow {}_RM \xrightarrow{f} {}_RN \xrightarrow{g} {}_RL \longrightarrow 0$$

in R-$\mathcal{M}od$ and we need to show that (a) f_* is a monomorphism, and (b) $\mathrm{im}(f_*) = \ker(g_*)$; then the sequence

$$0 \longrightarrow \mathrm{Hom}_R\,({}_R U,\,{}_R M) \xrightarrow{\ f_*\ } \mathrm{Hom}_R\,({}_R U,\,{}_R N) \xrightarrow{\ g_*\ } \mathrm{Hom}_R\,({}_R U,\,{}_R L)$$

will be exact.

If $fh = 0$ for some $h \in \mathrm{Hom}_R\,({}_R U,\,{}_R M)$ then $h = 0$ as f is injective. Therefore, f_* is injective, thus a monomorphism. This proves (a).

In order to prove (b), take $h \in \mathrm{im}(f_*) \subseteq \mathrm{Hom}_R\,({}_R U,\,{}_R N)$ so that $h = fk$ for some $k \in \mathrm{Hom}_R\,({}_R U,\,{}_R M)$. Then $g_*(h) = gh = gfk = 0$ as $\mathrm{im}(f) \subseteq \ker(g)$. It follows that $h \in \ker(g_*)$ and thus $\mathrm{im}(f_*) \subseteq \ker(g_*)$. Now take $h \in \ker(g_*)$ so that $gh = 0$. Then $h(U) \subseteq \ker(g) = \mathrm{im}(f)$. As a result, for $u \in U$, there is $m \in M$ with $h(u) = f(m)$ and, since f is injective, we can define $k \in \mathrm{Hom}_R\,({}_R U,\,{}_R M)$ by $k(u) = m$. It follows that $h = fk \in \mathrm{im}(f_*)$, which proves (b).

In a similar vein, the contravariant Hom-functor $\mathrm{Hom}_R\,(-,\,{}_R U)$ is left exact but right exactness (of either functor) needs additional assumptions on ${}_R U$, see Exercise 6.4.8 in Chap. 6.

4.3   Exercises

Exercise 4.3.1 Let ${}_R M \in R$-$\mathcal{M}od$ for a unital ring R be Noetherian. Let $f \in \mathrm{End}_R\,({}_R M)$. Show that f is injective provided it is surjective.

Exercise 4.3.2 Prove that every left Noetherian unital ring R is *von Neumann finite*, that is, for $x, y \in R$, $xy = 1$ implies that $yx = 1$.

Exercise 4.3.3 Let R be a commutative unital ring. Using Zorn's lemma, show that R is Noetherian if every prime ideal of R is finitely generated.

Exercise 4.3.4 Prove that, if R is a left Noetherian ring, then $M_n(R)$, $n \in \mathbb{N}$ is a left Noetherian ring.

Exercise 4.3.5 Provide an alternative proof of Corollary 4.1.2 by showing that a module ${}_R M$ has all submodules finitely generated (i.e., is Noetherian) if ${}_R M$ contains a Noetherian submodule ${}_R N$ such that ${}_R M / {}_R N$ is Noetherian. (That is, work only with condition (d) in Theorem 4.1. To this end, use Exercise 2.7.4 and an appropriate isomorphism theorem.)

Exercise 4.3.6 Let $_RM \in R\text{-}\mathcal{M}od$ for a unital ring R. Show that the following two conditions on $_RM$ are equivalent.

(a) $_RM$ is finitely generated;
(b) for every family $\{_RM_i \mid i \in I\} \subseteq R\text{-}\mathcal{M}od$ and each epimorphism
$f\colon \bigoplus_{i \in I} {_RM_i} \to {_RM}$ there exists a finite subset $J \subseteq I$ such that the composition

$$\bigoplus_{j \in J} {_RM_j} \xrightarrow{\iota_J} \bigoplus_{i \in I} {_RM_i} \xrightarrow{f} {_RM}$$

is surjective (where ι_J denotes the canonical monomorphism).

Exercise 4.3.7 Let $_RM \in R\text{-}\mathcal{M}od$ for a unital ring R. Show that $_RM$ is finitely generated if and only if, for every family $\{_RN_i \mid i \in I\}$ of submodules $_RN_i$ of $_RM$ such that $\langle \bigcup_{i \in I} {_RN_i} \rangle = {_RM}$, there exists a finite subset $J \subseteq I$ such that $\langle \bigcup_{j \in J} {_RN_j} \rangle = {_RM}$.

Exercise 4.3.8 Verify the properties of the zero morphism as stated in Sect. 4.2.1.

Exercise 4.3.9 Prove that the kernel and the cokernel of a morphism in a category with zero are unique up to isomorphism, provided they exist.

Exercise 4.3.10 Let $\mathscr{C}$ be an exact category. Show that for every $f \in \text{mor}(\mathscr{C})$ the following identities hold:

$$\ker(\text{cok}(\ker(f))) = \ker(f) \quad \text{and} \quad \text{cok}(\ker(\text{cok}(f))) = \text{cok}(f).$$

Exercise 4.3.11 Let $f, g \in \text{mor}(\mathscr{C})$ be "parallel arrows", that is, both f and g belong to $\text{Mor}_{\mathscr{C}}(A, B)$ for two objects A, B in $\mathscr{C}$. An *equalizer* of (f, g) is a morphism $e \in \text{Mor}_{\mathscr{C}}(C, A)$ such that $fe = ge$ and for every $h \in \text{Mor}_{\mathscr{C}}(D, A)$ with $fh = gh$ there is a unique $h' \in \text{Mor}_{\mathscr{C}}(D, C)$ such that $eh' = h$. Show that in $R\text{-}\mathcal{M}od$ every equalizer is a kernel.

Exercise 4.3.12 Identify the kernel and the cokernel of a morphism $f \in \mathscr{B}an_1$.

Exercise 4.3.13 Let $_RM$ and $_RN$ be left R-modules. Let $f\colon {_RM} \to {_RL}$ and $g\colon {_RN} \to {_RL}$ be module maps with the same codomain.

(i) Show that

$$_RM \times_L {_RN} = \{(x, y) \in {_RM} \times {_RN} \mid f(x) = g(y)\}$$

is a submodule of the product module $_RM \times {_RN}$.

(ii) Let $\pi_1: {}_RM \times {}_RN \to {}_RM$ and $\pi_2: {}_RM \times {}_RN \to {}_RN$ be the projections onto the first and the second coordinate, respectively. Show that the diagram below is commutative:

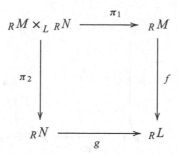

(iii) Prove that ${}_RM \times_L {}_RN$ has the universal property of a *pullback*; that is, given another commutative diagram of module maps of the form

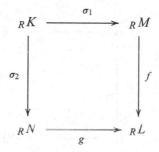

for some ${}_RK \in R\text{-}\mathcal{M}od$ there is a unique module map $h: {}_RK \to {}_RM \times_L {}_RN$ such that $\pi_1 \circ h = \sigma_1$ and $\pi_2 \circ h = \sigma_2$.

Exercise 4.3.14 Let ${}_RM$ and ${}_RN$ be left R-modules. Let $f: {}_RL \to {}_RM$ and $g: {}_RL \to {}_RN$ be module maps with the same domain.

(i) Show that

$$_RT = \{(f(x), -g(x)) \mid x \in {}_RL\}$$

is a submodule of the sum ${}_RM \oplus {}_RN$.

(ii) Let $_RM \oplus_T {}_RN$ denote the quotient module $(_RM \oplus {}_RN)/_RT$. Put $\alpha(y) = (y, 0) + {}_RT$, $y \in {}_RM$ and $\beta(z) = (0, z) + {}_RT$, $z \in {}_RN$. Show that the diagram below is commutative

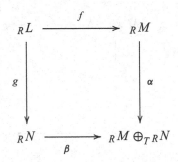

(iii) Prove that $_RM \oplus_T {}_RN$ has the universal property of a *pushout*; that is, given another commutative diagram of module maps of the form

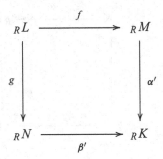

for some $_RK \in R\text{-}\mathcal{M}od$ there is a unique module map $h\colon {}_RM \oplus_T {}_RN \to {}_RK$ such that $h \circ \alpha = \alpha'$ and $h \circ \beta = \beta'$.

Artinian Modules

5

The dual condition to "ACC" is "DCC", the descending chain condition which we shall discuss in the present chapter. The resulting modules are termed *Artinian* after Emil Artin (1898–1962).

Historical Note Born in 1898 in Vienna (Austria), Emil Artin became one of the leading mathematicians of the twentieth century. He made major contributions to algebraic number theory, notably class field theory, abstract algebra and braid theory. In 1927 he solved Hilbert's 17th problem.

Artin's university studies at Vienna were interrupted when he was drafted in June 1918 but he continued them in Leipzig from 1919 onward where he was awarded his PhD in 1921. After a 1-year postdoctoral position at Göttingen during which period he worked closely with Emmy Noether and Helmut Hasse, Artin moved to Hamburg where he advanced to the rank of *Privatdozent*, having completed his Habilitation in 1923. In 1925 he accepted a position as associate professor at Hamburg University and was promoted to full professor in 1926.

Because of his wife's Jewish father and his own distaste for the Hitler regime, Artin lost his position in Hamburg and was forced to emigrate to the US where he got a professorship at Notre Dame University, Indiana in 1937. In 1938 Artin and his family moved again, to Indiana University in Bloomington. Apart from various visiting positions at Stanford, Ann Arbor, Boulder as well as in Japan, Artin spent the years between 1946 and 1957 at Princeton, which had become the mecca of mathematics during those years. Finally, in 1958, he returned to Hamburg where he would remain until he died of a heart attack in December 1962.

A detailed account of Emil Artin's life and work, written by J. J. O'Connor and E. F. Robinson can be found at
https://mathshistory.st-andrews.ac.uk/Biographies/Artin/

Interestingly enough there is no characterisation of Artinian modules in terms of elements, the dual concept to "finitely generated" turns out to be formulated entirely

© The Author(s), under exclusive license to Springer Nature Switzerland AG 2022   47
M. Mathieu, *Classically Semisimple Rings*,
https://doi.org/10.1007/978-3-031-14209-3_5

in categorical language so we need to introduce this notion first. We shall continue
our discussion of module-like categories in Sect. 5.4 below with a brief introduction
to abelian categories.

5.1   Finitely Cogenerated Modules

We first introduce the main concept of this chapter.

Definition 5.1.1 Let R be a unital ring. Let $_RM \in R\text{-}\mathcal{Mod}$. We say $_RM$ is *Artinian*
if it satisfies DCC, that is, every descending chain

$$_RN_1 \supseteq {_RN_2} \supseteq \ldots \supseteq {_RN_k} \supseteq \ldots$$

of submodules of $_RM$ becomes stationary, i.e., $_RN_k = {_RN_m}$ for all $k \geq m$ for some
$m \in \mathbb{N}$. The ring R is called *left Artinian* if $_RR$ is Artinian.

There is of course an analogous notion of Artinian right modules and right
Artinian rings, which is independent from left Artinian. A unital ring which is both
left and right Artinian is called *Artinian*.

In order to obtain a similar characterisation of Artinian modules as in Theorem 4.1, we need to introduce the following property. The reader may want to
compare this definition with Exercise 4.3.6.

Definition 5.1.2 Let R be a unital ring, and let $_RM \in R\text{-}\mathcal{Mod}$. We say $_RM$
is *finitely cogenerated* if, for every family $\{_RM_i \mid i \in I\} \subseteq R\text{-}\mathcal{Mod}$ and each
monomorphism $g \colon {_RM} \to \prod_{i \in I} {_RM_i}$, there exists a finite subset $J \subseteq I$ such that
the composition

$$_RM \xrightarrow{g} \prod_{i \in I} {_RM_i} \xrightarrow{p_J} \prod_{j \in J} {_RM_j}$$

is injective (where p_J is the canonical projection).

Exercises 4.3.7 and 5.5.1 illustrate nicely the interplay between the concepts of
finitely generated and finitely cogenerated modules.

We now come to the promised characterisation of Artinian modules.

Theorem 5.1.3 *Let $_RM \in R\text{-}\mathcal{Mod}$ for a unital ring R. The following conditions
are equivalent.*

(a) *$_RM$ is Artinian.*
(b) *Each strictly descending chain $_RN_1 \supset {_RN_2} \supset \ldots \supset {_RN_k} \supset \ldots$ of submodules
    $_RN_i \leq {_RM}$ is finite.*

(c) *Every non-empty set of submodules of $_RM$ contains a minimal element.*
(d) *Every quotient module of $_RM$ is finitely cogenerated.*

Proof The equivalence of the conditions (a) and (b) is evident.

(b) $\Rightarrow$ (c) Let $\mathscr{N}$ be a non-empty set of submodules of $_RM$ without a minimal element. Take $_RN_1 \in \mathscr{N}$; then there is $_RN_2 \in \mathscr{N}$ with $_RN_1 \supset {}_RN_2$. Let $_RN_1, \ldots, {}_RN_k \in \mathscr{N}$ be chosen such that $_RN_1 \supset {}_RN_2 \supset \ldots \supset {}_RN_k$. Since $_RN_k$ is not minimal in $\mathscr{N}$, there is $_RN_{k+1} \in \mathscr{N}$ strictly contained in $_RN_k$. By induction, we obtain an infinite strictly descending chain of submodules of $_RM$. Thus (b) fails.

(c) $\Rightarrow$ (d) Let $\{_RN_i \mid i \in I\}$ be a family of submodules of $_RM$. We put $\mathscr{N} = \{\bigcap_{j \in J} {}_RN_j \mid J \subseteq I \text{ finite}\}$, the set of all finite intersections of modules in $\{_RN_i \mid i \in I\}$. Clearly $\mathscr{N} \neq \emptyset$. Let $_RN_{\min}$ be a minimal element in $\mathscr{N}$. For each $_RN \in \mathscr{N}$, $_RN \cap {}_RN_{\min} \in \mathscr{N}$; hence $_RN \cap {}_RN_{\min} = {}_RN_{\min}$. It follows that $_RN_{\min}$ is the smallest element in $\mathscr{N}$. Since

$$\bigcap_{i \in I} {}_RN_i = \bigcap_{J \subseteq I \text{ finite}} \bigcap_{j \in J} {}_RN_j,$$

$\bigcap_{i \in I} {}_RN_i = 0$ implies that $\bigcap \mathscr{N} = 0$ and therefore $_RN_{\min} = 0$. By Exercise 5.5.1 below, $_RM$ is finitely cogenerated.

Every homomorphic image of $_RM$, and hence every quotient of $_RM$, inherits property (c). Thus, these quotient modules are finitely cogenerated too, proving (d).

(d) $\Rightarrow$ (a) Let $_RN_1 \supseteq {}_RN_2 \supseteq \ldots \supseteq {}_RN_k \supseteq \ldots$ be a descending chain of submodules of $_RM$. Put $_RN = \bigcap_{k \in \mathbb{N}} {}_RN_k$. By hypothesis, $_RM/{}_RN$ is finitely cogenerated. Since $\bigcap_{k \in \mathbb{N}} {}_RN_k/{}_RN = 0$, condition (d) together with Exercise 5.5.1 below imply that there is a finite subset $J \subseteq \mathbb{N}$ such that $\bigcap_{j \in J} {}_RN_j/{}_RN = 0$. Since $_RN_1/{}_RN \supseteq {}_RN_2/{}_RN \supseteq \ldots \supseteq {}_RN_k/{}_RN \supseteq \ldots$ it follows that $_RN_r/{}_RN = 0$ for some $r \in \mathbb{N}$, that is, $_RN_r = {}_RN$. As a result, $_RN_{r+k} = {}_RN_r$ for all $k \in \mathbb{N}$ wherefore $_RM$ is Artinian. $\square$

The proof of the next result is analogous to the one of Theorem 4.1.1 and is therefore left to the reader, see Exercise 5.5.2.

Theorem 5.1.4 *Let R be a unital ring. Let*

$$0 \longrightarrow {}_RM_1 \overset{f}{\longrightarrow} {}_RM_2 \overset{g}{\longrightarrow} {}_RM_3 \longrightarrow 0$$

be a short exact sequence in R-$\mathscr{M}od$. Then $_RM_2$ is Artinian if and only if both $_RM_1$ and $_RM_3$ are Artinian.

Unsurprisingly, similar consequences can be drawn from Theorem 5.1.4 as in the Noetherian case. The proofs are identical to that situation.

Corollary 5.1.5 *Let R be a unital ring. Let $_RN$ be a submodule of $_RM \in R\text{-}\mathcal{Mod}$. Then $_RM$ is Artinian if and only if both $_RN$ and $_RM/_RN$ are Artinian.*

Corollary 5.1.6 *Every finite direct sum of Artinian modules is Artinian.*

We have the following characterisation of left Artinian rings.

Corollary 5.1.7 *For every unital ring R, the following conditions are equivalent.*

(a) *R is left Artinian;*
(b) *Every finitely generated unital left R-module is finitely cogenerated.*

Proof (b) $\Rightarrow$ (a) As R is unital, $_RR$ is finitely generated. It follows that every quotient module $_RR/_RL$, where L is a proper left ideal of R, is finitely generated as well and, by assumption, finitely cogenerated. Therefore, by Theorem 5.1.3, $_RR$ is Artinian.

(a) $\Rightarrow$ (b) Let $_RM \in R\text{-}\mathcal{Mod}$. If $_RM$ is finitely generated, then $_RM$ is a homomorphic image of a free finitely generated left R-module $_RN$. As $_RN$ is (isomorphic to) a finite direct sum of $_RR$ (Exercise 2.7.6), Corollary 5.1.6 entails that $_RN$ is Artinian. Theorem 5.1.3 thus yields the statement.   □

As a consequence, a unital ring R is left Artinian if and only if every module in $R\text{-}\mathcal{mod}$ is Artinian, another parallel to the situation of Noetherian rings.

We shall now start to explore the differences between the theories of Noetherian and of Artinian rings. For a start, it is very easy to give an example of a Noetherian ring which is not Artinian: $\mathbb{Z}$ is, as a principal ideal domain, clearly a Noetherian ring but it is not Artinian; for instance, the descending chain

$$\mathbb{Z} \supset 2\mathbb{Z} \supset 2^2\mathbb{Z} \supset \ldots \supset 2^k\mathbb{Z} \supset \ldots$$

is infinite.

5.2   Commutative Artinian Rings

In this section we want to study some special properties of Artinian rings that are commutative.

Definition 5.2.1 Let R be a principal ideal domain. A module $_RM \in R\text{-}\mathcal{Mod}$ is said to be a *torsion module* if $\operatorname{Ann}_R(m) \neq 0$ for every $m \in M$.

Proposition 5.2.2 *Every finitely generated torsion module over a principal ideal domain is Artinian.*

Proof Let R be a principal ideal domain, and let $_R M \in R\text{-}mod$ with generators $m_1, \ldots, m_k$. Suppose that $_R M$ is a torsion module. Then, by Exercise 2.7.9,

$$_R M = R m_1 + \ldots + R m_k \quad \text{and} \quad R m_i \cong R/\text{Ann}_R(m_i), \ 1 \leq i \leq k.$$

It follows that $_R M$ is a homomorphic image of $\bigoplus_{i=1}^k R/\text{Ann}_R(m_i)$; thus, by Corollaries 5.1.6 and 5.1.5, it suffices to show that each summand $R/\text{Ann}_R(m_i)$ is Artinian.

In order to prove this claim, take a non-zero ideal I of R (this is where we use $\text{Ann}_R(m_i) \neq 0$ for each i). By hypothesis, $I = aR$ for some $a \in R$. We will show that the ideal lattice of R/I is finite; thus DCC on submodules of R/I certainly must hold.

As R is a principal ideal domain, it is a unique factorisation domain. As a result, we can write a in the form $a = p_1^{n_1} p_2^{n_2} \cdots p_\ell^{n_\ell}$ with irreducible elements $p_j \in R$ and $n_1, \ldots, n_\ell \in \mathbb{N}$. For $J = bR$, $b \in R$ we have $I \subseteq J$ if and only if $b \mid a$. The uniqueness in the factorisation of a implies that there are only finitely many divisors of a; hence only finitely many ideals in R containing I. Consequently, R/I has at most finitely many ideals. $\square$

The reader may want to compare the above result with the fact that *every* finitely generated module over a principal ideal domain is Noetherian.

Commutative Artinian rings are in a way the simplest rings after fields. To give evidence to that fact, we record a few more of their nice properties (some of which do extend appropriately to non-commutative Artinian rings).

Definition 5.2.3 Let R be a commutative unital ring. Then

$$\text{rad}(R) = \bigcap \{I \mid I \text{ maximal ideal in } R\}$$

and

$$\text{nil}(R) = \bigcap \{P \mid P \text{ prime ideal in } R\}$$

are called the *Jacobson radical* and the *nil radical* of R, respectively.

By definition, both $\text{nil}(R)$ and $\text{rad}(R)$ are ideals of R and, since every maximal ideal is a prime ideal—the quotient by it is a field, hence an integral domain—, we have $\text{nil}(R) \subseteq \text{rad}(R)$. In general, this inclusion will be strict.

Proposition 5.2.4 *In an Artinian commutative ring R, every prime ideal is maximal. Hence* $\text{nil}(R) = \text{rad}(R)$.

Proof Let P be a prime ideal of R. Then $S = R/P$ is an Artinian integral domain (Corollary 5.1.5). Let $x \in S \setminus \{0\}$. Let us denote the ideal generated by $a \in R$ by (a). Since $(x^n) \supseteq (x^{n+1})$ for all $n \in \mathbb{N}$, by DCC there is some $m \in \mathbb{N}$ such that

$(x^m) = (x^{m+1})$; that is, $x^m = x^{m+1}y$ for some $y \in S$. Since S is an integral domain, we find that $1 = xy$, in other words, x is invertible. It follows that every non-zero element in S is invertible so that S is a field. As a result, P is a maximal ideal. $\square$

Proposition 5.2.5 *In an Artinian commutative ring R, the nil radical $\mathrm{nil}(R)$ is nilpotent, that is, $\mathrm{nil}(R)^k = 0$ for some $k \in \mathbb{N}$.*

Proof Since $\mathrm{nil}(R)^n \supseteq \mathrm{nil}(R)^{n+1}$ for all $n \in \mathbb{N}$, DCC gives $\mathrm{nil}(R)^n = \mathrm{nil}(R)^k$ for some $k \in \mathbb{N}$ and all $n \geq k$. Put $I = \mathrm{nil}(R)^k$. Suppose $I \neq 0$; then the set $\mathscr{J}$ of all ideals J in R with $IJ \neq 0$ is non-empty. Since R is Artinian, there is a minimal element J_0 in $\mathscr{J}$. Take $x \in J_0$ with $xI \neq 0$. As $(x) \subseteq J_0$ we conclude that $(x) = J_0$. Since $(xI)I = xI^2 = xI \neq 0$, an analogous argument yields $xI = (x) = J_0$. It follows that $x = xy$ for some $y \in I$ from which we obtain

$$x = xy = xy^2 = \ldots = xy^n \quad \text{for all } n \in \mathbb{N}.$$

Since $y \in I = \mathrm{nil}(R)^k \subseteq \mathrm{nil}(R)$ and every element in $\mathrm{nil}(R)$ is nilpotent (Exercise 5.5.5 below), $y^\ell = 0$ for some $\ell \in \mathbb{N}$. This implies that $x = xy^\ell = 0$ contradicting the choice of x. We conclude that $I = 0$ as required. $\square$

5.3   Artinian vs. Noetherian Modules

We already observed that the ring $\mathbb{Z}$ is Noetherian but not Artinian. Here is another example of the same ilk.

Example 5.3.1 We shall use again the ring from Example 4.3. The ring

$$R = \mathbb{Z}\langle s, t \mid t^2 = ts = 0 \rangle = \mathbb{Z}[s] \oplus \mathbb{Z}[s]t$$

is left Noetherian (but not right Noetherian). Moreover, R is neither left nor right Artinian since the module $\mathbb{Z}[s] \cong R/\mathbb{Z}[s]t$ is not Artinian; it contains $(\mathbb{Z}[s]s^k)_{k \in \mathbb{N}}$ as an infinite strictly descending chain of submodules.

In the opposite direction, there is the following surprising result for rings (it does not extend to modules, though).

5.3.2 Hopkins–Levitzki Theorem *Every left Artinian ring is left Noetherian.*

We defer the proof until Chap. 7 but the reader can already try their hands at the commutative case, see Exercise 5.5.8.

Of course, an analogous statement holds with "left" replaced by "right" in both the assumption and the conclusion.

We conclude our discussion by bringing the theories of Artinian and Noetherian modules close together once more. The following result should remind the reader of some facts they learned in Linear Algebra.

Proposition 5.3.3 *Let R be a unital ring, and let $_RM \in R\text{-}\mathcal{Mod}$. For every $f \in \text{End}_R(_RM)$, the following holds.*

(i) *If $_RM$ is Artinian then there is $n \in \mathbb{N}$ such that*

$$\ker(f^n) + \text{im}(f^n) = {}_RM;$$

in particular, f is surjective if it is injective.

(ii) *If $_RM$ is Noetherian then there is $n \in \mathbb{N}$ such that*

$$\ker(f^n) \cap \text{im}(f^n) = 0;$$

in particular, f is injective if it is surjective.

(iii) *If $_RM$ is both Artinian and Noetherian then there is $n \in \mathbb{N}$ such that*

$$\ker(f^n) \oplus \text{im}(f^n) = {}_RM.$$

In addition, the following conditions are equivalent:
(a) *f is injective;*
(b) *f is surjective;*
(c) *f is bijective.*

The third statement in the above proposition is also known as "Fitting's Lemma". In all statements, f^n stands for the n-th iterate of the mapping f.

Proof

(i) Suppose $_RM$ is Artinian. Then the descending chain of submodules

$$\text{im}(f) \supseteq \text{im}(f^2) \supseteq \ldots \supseteq \text{im}(f^n) \supseteq \ldots$$

becomes stationary; that is, $\text{im}(f^{n+k}) = \text{im}(f^n)$ for some $n \in \mathbb{N}$ and all $k \in \mathbb{N}$. Therefore, for given $x \in {}_RM$, there is $y \in {}_RM$ such that $f^n(x) = f^{2n}(y)$; consequently, $x - f^n(y) \in \ker(f^n)$. It follows that $x = x - f^n(y) + f^n(y) \in \ker(f^n) + \text{im}(f^n)$, as claimed.

If $\ker(f) = 0$ then $\ker(f^n) = 0$; thus $\text{im}(f^n) = {}_RM$ which implies $\text{im}(f) = {}_RM$.

(ii) Suppose $_R M$ is Noetherian. Then the ascending chain of submodules

$$\ker(f) \subseteq \ker(f^2) \subseteq \dots \subseteq \ker(f^n) \subseteq \dots$$

becomes stationary; that is, $\ker(f^{n+k}) = \ker(f^n)$ for some $n \in \mathbb{N}$ and all $k \in \mathbb{N}$. In particular, $\ker(f^{2n}) = \ker(f^n)$. Thus, for $x \in \ker(f^n) \cap \operatorname{im}(f^n)$, say $x = f^n(y)$, $y \in {}_R M$, we have $f^{2n}(y) = f^n(x) = 0$. As $y \in \ker(f^{2n}) = \ker(f^n)$, we obtain $x = 0$ as claimed.

If $\operatorname{im}(f) = {}_R M$ then $\operatorname{im}(f^n) = {}_R M$; thus $\ker(f^n) = 0$ which implies that $\ker(f) = 0$.

(iii) is a direct consequence of (i) and (ii). The proof is complete.

$\square$

5.4   Abelian Categories

Let $\mathscr{C}$ be a category which is both additive and exact; see p. 23 and 42, respectively. Then $\mathscr{C}$ is called *abelian*. Abelian categories were introduced by A. Grothendieck in 1957 as an abstract setting which captured many features of module categories. In this section we aim to illustrate this fact by discussing various results familiar from module theory that can be obtained without the use of elements in the objects.

Proposition 5.4.1 *Let $f \in \operatorname{Mor}_{\mathscr{C}}(A, B)$, where $\mathscr{C}$ is an abelian category. Then f is an isomorphism if and only if it is both a monomorphism and an epimorphism.*

Proof The "only if"-part follows directly from the definitions; see Definition 1.14. Suppose f is a monomorphism. If $fg = 0$ for some $g \in \operatorname{Mor}_{\mathscr{C}}(D, A)$ then $g = 0$. Therefore g can be uniquely factored through $0 \longrightarrow A$, that is, $\ker(f) = 0$. Note that $\operatorname{cok}(\ker(f)) = \operatorname{cok}(0) = 1_A$; thus we have a commutative diagram

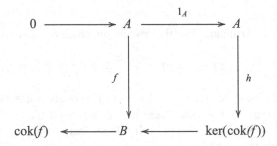

with a unique morphism $h \in \operatorname{Mor}_{\mathscr{C}}(A, \ker(\operatorname{cok}(f)))$. Thus $f = \ker(\operatorname{cok}(f))$.

Suppose in addition that f is an epimorphism. Then $\mathrm{cok}(f)\, f = 0$ so $\mathrm{cok}(f) = 0$. Hence $\ker(\mathrm{cok}(f)) = 1_B$ and the above diagram becomes

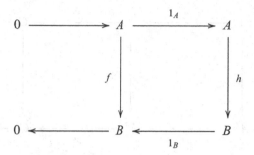

The two properties $f = \ker(\mathrm{cok}(f))$ and $f = \mathrm{cok}(\ker(f))$, respectively now give us the existence of unique morphisms k, $k' \in \mathrm{Mor}_{\mathscr{C}}(B, A)$, respectively making the diagrams below commutative, compare (4.2.1) and (4.2.2):

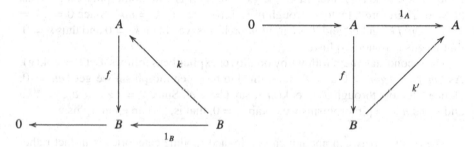

that is, $fk = 1_B$ and $k'f = 1_A$. Thus f is an isomorphism as claimed. □

Corollary 5.4.2 *A morphism $f \in \mathrm{Mor}_{\mathscr{C}}(A, B)$, where $\mathscr{C}$ is an abelian category, is an epimorphism if and only if $\ker(\mathrm{cok}(f)) = B$.*

Proof In the proof of the above proposition we observed that f is an epimorphism if and only if $B = \mathrm{cok}(\ker(f))$. Since $\ker(\mathrm{cok}(f)) \cong \mathrm{cok}(\ker(f))$, see Exercise 5.5.10, we obtain the claim. □

A typical technique to obtain results in abelian categories that are analogous to those in module categories, such as the isomorphism theorems for example (see also Exercise 5.5.10 below), is the use of diagram lemmas. The difference to working in a concrete category lies in the fact that, in a general abelian category, there are no elements to chase around. Hence the arguments become somewhat lengthy; for that reason we confine ourselves to the sample below, and leave the more elaborate results, such as the 5−Lemma, the 3 × 3−Lemma or the Snake Lemma, to more comprehensive expositions (see [30], e.g.).

5.4.3 Short Five Lemma *Suppose we are given the following commutative diagram in the abelian category $\mathscr{C}$:*

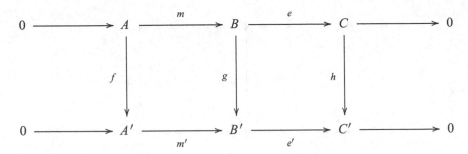

Suppose both rows are exact and the morphisms f and h are both monomorphisms (epimorphisms). Then g is a monomorphism (an epimorphism).

Proof Let $k = \ker(g)$; then $hek = e'gk = 0$. As h is a monomorphism this yields $ek = 0$. Therefore k factors through $m = \ker(e)$, say as $k = mk'$. Since $0 = gk = gmk' = m'fk'$ and m' and f are monomorphisms we obtain $k' = 0$ and thus $k = 0$, that is, g is a monomorphism.

The second statement follows by duality, or explicitly as follows. Let $\ell = \mathrm{cok}(g)$. As $\ell m'f = \ell gm = 0$ and f is assumed to be an epimorphism we get $\ell m' = 0$. Hence ℓ factors through $e' = \mathrm{cok}(m')$, say $\ell = \ell'e'$. Since $0 = \ell g = \ell'e'g = \ell'he$ and e and h are epimorphisms we obtain $\ell = 0$, that is, g is an epimorphism. $\square$

The relation between abelian categories and module categories is in fact rather intimate. The Freyd–Mitchell theorem states that for every small abelian category there is an full exact embedding in $R\text{-}\mathscr{M}od$ for a suitable ring R. See, e.g., Sect. 4.14 in [30].

There exist various alternative but equivalent ways to define an abelian category, some even derive the abelian group structure on the Hom-sets from other basic axioms; see [17, Chap. 2] and [30, Chap. 4]. A very readable account is given in Chap. 7 of [29].

5.5   Exercises

Exercise 5.5.1 Let $_RM \in R\text{-}\mathscr{M}od$ for a unital ring R. Show that $_RM$ is finitely cogenerated if and only if, for every family $\{_RN_i \mid i \in I\}$ of submodules $_RN_i$ of $_RM$ such that $\bigcap_{i \in I} {}_RN_i = 0$, there exists a finite subset $J \subseteq I$ such that $\bigcap_{j \in J} {}_RN_j = 0$.

Exercise 5.5.2 Write out the details of the proof of Theorem 5.1.4.

Exercise 5.5.3 Let R be an integral domain, and let $_RM \in R\text{-}\mathcal{M}od$. The *torsion submodule* of $_RM$ is defined as

$$\text{tor}(_RM) = \{m \in M \mid \text{Ann}_R(m) \neq 0\}.$$

Prove the following statements.

(i) $\text{tor}(_RM) \leq {}_RM$;
(ii) $\text{tor}(_RM/\text{tor}(_RM)) = 0$;
(iii) for $f \in \text{Hom}_R({}_RM, {}_RN)$, where $_RN \in R\text{-}\mathcal{M}od$, we have $f(\text{tor}(_RM)) \subseteq \text{tor}(_RN)$.

Exercise 5.5.4 In an Artinian commutative ring R, there are only finitely many prime ideals.

Exercise 5.5.5 Let R be a commutative unital ring. Show that the definitions of the nil radical of R given in Definition 5.2.3 and in Exercise 1.3.10 agree with each other.

Exercise 5.5.6 Let K be a field and let V be a K-vector space. Show that the following three conditions are equivalent.

(a) V is finite dimensional;
(b) V satisfies ACC;
(c) V satisfies DCC.

Exercise 5.5.7 Let R be a commutative unital ring and let $M_1, \ldots, M_\ell$ be maximal ideals in R such that $M_1 M_2 \cdots M_\ell = 0$. Show that R is Noetherian if and only if R is Artinian.

Exercise 5.5.8 Show that an Artinian commutative ring R is Noetherian (compare Theorem 5.3.2) by employing Propositions 5.2.4 and 5.2.5 together with Exercises 5.5.4 and 5.5.7.

Exercise 5.5.9 Let $\mathscr{C}$ be an abelian category. Show that every morphism $f \in \text{mor}(\mathscr{C})$ has a factorisation $f = me$, where $m = \ker(\text{cok}(f))$ and $e = \text{cok}(\ker(f))$ (In particular, m is a monomorphism and e is an epimorphism. This is the analogue of the canonical factorisation of module maps (Theorem 3.1.1).)

The next exercise is the analogue of the First Isomorphism Theorem in module theory (Theorem 3.1.2) in abelian categories.

Exercise 5.5.10 Let $\mathscr{C}$ be an abelian category. For each $f \in \mathrm{mor}(\mathscr{C})$, there is a canonically associated isomorphism $h \colon \mathrm{cok}(\ker(f)) \longrightarrow \ker(\mathrm{cok}(f))$ which arises as follows. Starting with the diagram

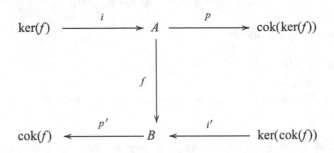

for a given $f \in \mathrm{Mor}_{\mathscr{C}}(A, B)$, use the definition of kernel and cokernel in Sect. 4.2.1 in order to obtain unique morphisms $g \colon A \longrightarrow \ker(\mathrm{cok}(f))$ and $k \colon \mathrm{cok}(\ker(f)) \longrightarrow B$, respectively with the properties $f = i'g$ and $f = kp$, respectively. In the commutative diagram

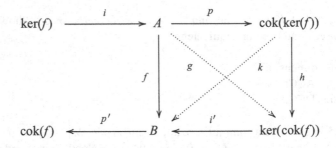

g may be uniquely factored through $\mathrm{cok}(\ker(f))$ by h because $0 = fi = i'gi$, hence $0 = gi$ as i' is a monomorphism. Likewise k factors through $\ker(\mathrm{cok}(f))$ via h'. Using $i'g = f = kp$ conclude that $h = h'$ and finally employ Proposition 5.4.1 to prove that h is an isomorphism.

Exercise 5.5.11 Let R be a left Noetherian ring. Show that $R\text{-}mod$ is an abelian category.

Exercise 5.5.12 Repeat Exercise 4.3.11 for an abelian category $\mathscr{C}$ in place of $R\text{-}\mathcal{M}od$.

Exercise 5.5.13 Fill in the details in the following statement: The additive category $\mathscr{B}an_\infty$ is not abelian since the inclusion $\ell^1 \to c_0$ is both a monomorphism and an epimorphism but not an isomorphism. (For the definition of the spaces, see p. 68.)

Simple and Semisimple Modules

6

In this chapter, which is at the very heart of our exposition, we shall apply module theory to study the structure of a certain class of rings; these will be termed (*classically*) *semisimple*. To this end, we will, once again, introduce some special classes of modules and investigate their properties.

From this point on, we shall only consider unital rings (but still state this explicitly) and unital modules. These were introduced in the Exercises in Chap. 1 where it was also shown that this is no restriction of the generality.

6.1 Decomposition of Modules

We turn our attention to the modules that are of the utmost interest to us in this book.

Definition 6.1.1 Let R be a unital ring, and let $_R M \in R\text{-}\mathcal{M}od$. Then

(i) $_R M$ is called *simple* (or, *irreducible*) if $_R M \neq 0$ and $_R M$ contains no non-trivial submodules; that is, 0 and $_R M$ are the only submodules of $_R M$.

(ii) $_R M$ is called *semisimple* (or, *completely reducible*) if every submodule of $_R M$ is a direct summand; that is, for each $_R N \leq {}_R M$ there is $_R N' \leq {}_R M$ such that $_R M = {}_R N \oplus {}_R N'$.

Remark 6.1.2 Identifying $_R M \in R\text{-}\mathcal{M}od$ with the associated representation $R \to \text{End}(M)$ (compare Sect. 1.1.4), simplicity of $_R M$ is the same as the property that there are no non-trivial subgroups of $(M, +)$ which are invariant under the R-action. This interpretation gives rise to the terminology 'irreducible'.

Examples 6.1.3

1. An abelian group G is simple as a group if and only if it is simple as a $\mathbb{Z}$-module. The maximal ideals of $\mathbb{Z}$ are of the form $p\mathbb{Z}$ for a prime number p. Hence, by Exercise 6.4.1 below, G is simple if and only if G is isomorphic to $\mathbb{Z}_p$.
2. Let V be a non-zero vector space over the field K. Put $R = \text{End}_K(V)$. Then $V \in R\text{-}\mathcal{M}od$ via $f \cdot v = f(v)$, where $f \in R$, $v \in V$. We show that $_RV$ is simple: take $v \in V \setminus \{0\}$ and let $w \in V$. Extend $\{v\}$ to a basis B of V and define $f \in \text{End}_K(V)$ by $f(v) = w$ and $f(b) = 0$ for each $b \in B \setminus \{v\}$. Then $f \cdot v = w$; consequently, Exercise 6.4.1 yields the result.

Our next result, though very easy to prove, is fundamental for the application of simple modules in ring theory. It was first obtained by I. Schur in 1905 in a different language (see Exercise 6.4.2 below).

6.1.4 Schur's Lemma *Let R be a unital ring, and let $_RM \in R\text{-}\mathcal{M}od$ be simple. Then $\text{End}_R(_RM)$ is a division ring.*

Proof Since $\ker(f) \leq {}_RM$ for each $f \in \text{End}_R(_RM)$, either $\ker(f) = 0$ — in which case f is injective — or $\ker(f) = {}_RM$, that is, $f = 0$. Similarly, $\text{im}(f) \leq {}_RM$ implies that $\text{im}(f) = {}_RM$ — in which case f is surjective — or $\text{im}(f) = 0$, that is $f = 0$. It follows that every non-zero endomorphism of $_RM$ is bijective, and hence $\text{End}_R(_RM)$ is a division ring.   □

Remark 6.1.5 Suppose $_RM$ is a simple module over a commutative unital ring R. Then $\text{End}_R(_RM)$ is in fact a field. By Schur's Lemma, all we need to show is that $\text{End}_R(_RM)$ is commutative. By Exercise 6.4.1, $_RM \cong R/I$ for a maximal ideal I of R. Since R/I is a commutative unital ring (in fact, a field), $R/I \cong \text{End}_{R/I}(R/I)$; compare Exercise 2.7.1. Putting this together we find

$$\text{End}_R(_RM) \cong \text{End}_{R/I}(R/I) \cong R/I$$

is commutative as $I = \text{Ann}_R(M)$.

We shall now study the relations between simple and semisimple modules. It is evident from the definition that every simple module is semisimple. To obtain information in the other direction, we first record the following useful observation.

Proposition 6.1.6 *Every submodule and every quotient of a semisimple module is semisimple.*

Proof Let $_RM \in R\text{-}\mathcal{M}od$ for a unital ring R be semisimple. Let $_RN \leq {}_RM$. Take $_RL \leq {}_RN$; by assumption, there is $_RL' \leq {}_RM$ such that $_RL \oplus {}_RL' = {}_RM$. Therefore, setting $_RL'' = {}_RN \cap {}_RL'$, we get

$$_RL \oplus {}_RL'' = {}_RL \oplus ({}_RN \cap {}_RL') = {}_RN \cap ({}_RL \oplus {}_RL') = {}_RN$$

by the modular law (Exercise 2.7.8). Hence $_RN$ is semisimple.

Let f be an R-module map defined on $_RM$ and let $_RL \leq \text{im}(f)$. Then $_RN = f^{-1}(_RL) \leq {}_RM$ and, by hypothesis, $_RN \oplus {}_RN' = {}_RM$ for some $_RN' \leq {}_RM$. Since $\ker(f) \subseteq {}_RN$, it follows that

$$\text{im}(f) = f(_RM) = f(_RN) \oplus f(_RN') = {}_RL \oplus f(_RN')$$

and thus $\text{im}(f)$ is semisimple. The first isomorphism theorem (Theorem 3.1.2) completes the argument. □

Lemma 6.1.7 *Let R be a unital ring. Every non-zero semisimple module $_RM \in R\text{-}\mathcal{M}od$ contains a simple submodule.*

Proof In case every submodule $_RN$ of $_RM$ is either 0 or $_RM$, there is nothing to prove. Thus we can take a non-zero submodule $_RN \leq {}_RM$ together with an element $m \in {}_RM \setminus {}_RN$ and, without loss of generality, we can assume that m generates $_RM$, cf. Proposition 6.1.6. Zorn's Lemma provides us with a maximal submodule $_RL$ of $_RM$ containing $_RN$ and not containing m. Indeed, the non-empty set $\mathcal{N} = \{_RK \leq {}_RM \mid {}_RN \subseteq {}_RK \text{ and } m \notin {}_RK\}$ is inductively ordered so has a maximal element $_RL$. By hypothesis, $_RM = {}_RL \oplus {}_RL'$ for some non-zero $_RL' \leq {}_RM$. We claim that $_RL'$ is simple. Suppose $_RL''$ is a non-zero submodule of $_RL'$. As $_RL$ is maximal in $\mathcal{N}$, $_RL \oplus {}_RL''$ must contain m and hence, $_RL \oplus {}_RL'' = {}_RM$. As a result, $_RL' \subseteq {}_RL''$ wherefore $_RL'' = {}_RL'$ and $_RL'$ is simple. □

With the above preparations at hand, we can now 'completely reduce' any semisimple module into its simple constituents.

Proposition 6.1.8 *The following conditions on $_RM \in R\text{-}\mathcal{M}od$ are equivalent:*

(a) *$_RM$ is semisimple;*
(b) *$_RM$ is the direct sum of simple submodules;*
(c) *$_RM$ is the sum of simple submodules.*

Proof By convention, $0 = \bigoplus_{i \in \emptyset} {}_RM_i$; thus we can without loss of generality assume that $_RM \neq 0$. Suppose $_RM = \sum_{i \in I} {}_RM_i$ where each $_RM_i \leq {}_RM$ is simple. Let $_RN \leq {}_RM$. Let $\mathbf{S}$ be the set of all subsets $J \subseteq I$ such that

(i) $\sum_{j \in J} {}_RM_j$ is a direct sum;   and   (ii) $_RN \cap \sum_{j \in J} {}_RM_j = 0$.

Clearly $\emptyset \in \mathbf{S}$ so $\mathbf{S} \neq \emptyset$. Since $\mathbf{S}$ is inductively ordered, Zorn's lemma provides us with a maximal element $J_{\max}$. By construction,

$$_R M' := {}_R N + \sum_{j \in J_{\max}} {}_R M_j = {}_R N \oplus \bigoplus_{j \in J_{\max}} {}_R M_j.$$

We will show that $_R M_i \subseteq {}_R M'$ for every $i \in I$ which yields $_R M = {}_R M'$. For each $i \in I$,

$$_R M' \cap {}_R M_i = {}_R M_i \quad \text{or} \quad {}_R M' \cap {}_R M_i = 0$$

as each $_R M_i$ is simple. The latter case implies $_R M' + {}_R M_i = {}_R N \oplus \bigoplus_{j \in J_{\max}} {}_R M_j \oplus {}_R M_i$ and hence, by maximality of $J_{\max}$, $i \in J_{\max}$. The first case also entails that $_R M_i \subseteq {}_R M'$ and therefore $_R M' = {}_R M$.

Choosing $_R N = 0$ in the above argument gives the implication (c) $\Rightarrow$ (b) and choosing $_R N$ arbitrary yields (b) $\Rightarrow$ (a).

(a) $\Rightarrow$ (c) Let $_R N$ be the sum of all simple submodules of $_R M$; then $_R N \neq 0$, by Lemma 6.1.7. Let $_R N' \leq {}_R M$ be such that $_R N \oplus {}_R N' = {}_R M$. By Proposition 6.1.6, $_R N'$ is semisimple, however it does not contain any non-zero simple submodule. By Lemma 6.1.7, $_R N' = 0$ and therefore $_R N = {}_R M$. This proves (c).   □

Proposition 6.1.9 *Suppose that $_R M \in R\text{-}\mathcal{M}od$ is semisimple and that $_R M = \bigoplus_{i \in I} {}_R M_i$ is a decomposition of $_R M$ into simple submodules $_R M_i$. If $_R N \leq {}_R M$ is simple then $_R N \cong {}_R M_j$ for some $j \in I$.*

Proof As $_R M = \bigoplus_{i \in I} {}_R M_i$, we may, and will, consider $_R M$ as a submodule of $\prod_{i \in I} {}_R M_i$. Let $p_j : \prod_{i \in I} {}_R M_i \to {}_R M_j$ be the canonical projection onto the jth component. Since $_R N \neq 0$, $p_j({}_R N) \neq 0$ for at least one $j \in I$. Therefore, $f = p_{j|_{R N}} \in \text{Hom}_R({}_R N, {}_R M_j)$ is a non-zero homomorphism. Since $_R M_j$ is simple, $\text{im}(f) = {}_R M_j$; since $_R N$ is simple, $\ker(f) = 0$. It follows that f is an isomorphism.   □

It can be shown that the cardinality of the set I in a decomposition such as in Proposition 6.1.9 is unique; see Exercise 6.4.3 below.

A neat criterion to test a module for semisimplicity, which will turn out to be useful in Chap. 7, is the following.

Proposition 6.1.10 *Let $_R M \in R\text{-}\mathcal{M}od$ be finitely cogenerated. Then $_R M$ is semisimple if and only if the intersection over all maximal submodules of $_R M$ is zero.*

Proof Suppose first that $_R M \in R\text{-}\mathcal{M}od$ is semisimple and that $_R M = \bigoplus_{i \in I} {}_R M_i$ is a decomposition of $_R M$ into simple submodules $_R M_i$. For each $j \in I$, set $_R N_j = \bigoplus_{i \neq j} {}_R M_i$. Then $_R N_j$ is a maximal submodule of $_R M$ as $_R M / {}_R N_j \cong {}_R M_j$

which is simple. We have $\bigcap_{j \in I} {}_R N_j = 0$, hence the intersection over all maximal submodules of ${}_R M$ is zero.

Conversely, let $\{ {}_R N_i \mid i \in I \}$ be the collection of all maximal submodules of ${}_R M$. For each $j \in I$, put ${}_R M_i = {}_R M / {}_R N_i$ which is a simple left R-module. Suppose that $\bigcap_{i \in I} {}_R N_i = 0$. By Exercise 5.5.1, there is a finite subset $J \subseteq I$ such that $\bigcap_{j \in J} {}_R N_j = 0$. Consequently, the R-module map ${}_R M \longrightarrow \bigoplus_{j \in J} {}_R M_j$ is an embedding of ${}_R M$ into a semisimple module, hence ${}_R M$ is semisimple (Proposition 6.1.6). □

Our next immediate goal is to determine for which rings every (left) module is semisimple. Clearly, the question for which ring every module is *simple* does not make sense, since we can always form direct sums. The key to the answer to the above question is the following concept for short exact sequences of modules. (Compare with (3.3.2) on p. 32.)

Definition 6.1.11 Let $0 \longrightarrow {}_R M_1 \xrightarrow{f} {}_R M_2 \xrightarrow{g} {}_R M_3 \longrightarrow 0$ be a short exact sequence in $R\text{-}\mathcal{M}od$. We say the sequence *splits* if $\operatorname{im}(f) = \ker(g)$ is a direct summand in ${}_R M_2$.

Theorem 6.1.12 *The following conditions are equivalent for a unital ring R.*

(a) *Every short exact sequence in $R\text{-}\mathcal{M}od$ splits;*
(b) *Every ${}_R M \in R\text{-}\mathcal{M}od$ is semisimple;*
(c) *Every finitely generated ${}_R M \in R\text{-}\mathcal{M}od$ is semisimple;*
(d) *Every cyclic ${}_R M \in R\text{-}\mathcal{M}od$ is semisimple;*
(e) ${}_R R$ *is semisimple.*

Proof (a) $\Rightarrow$ (b) Let ${}_R N \leq {}_R M$ and consider the short exact sequence $0 \longrightarrow {}_R N \longrightarrow {}_R M \longrightarrow {}_R M / {}_R N \longrightarrow 0$. . By hypothesis, this sequence splits and hence ${}_R N$ is a direct summand in ${}_R M$.
(b) $\Rightarrow$ (c) $\Rightarrow$ (d) $\Rightarrow$ (e) are obviously valid.
(b) $\Rightarrow$ (a) If each submodule of a module in $R\text{-}\mathcal{M}od$ is a direct summand so in particular the image $0 \longrightarrow {}_R M_1 \longrightarrow {}_R M_2$ in a given short exact sequence as in Definition 6.1.11.
(e) $\Rightarrow$ (b) Let ${}_R M \in R\text{-}\mathcal{M}od$. For $m \in M$, the cyclic module Rm is semisimple since it is a homomorphic image of the semisimple module ${}_R R$ (Proposition 6.1.6). By Proposition 6.1.8, Rm is thus the sum $\sum_{j \in J_m} {}_R N_j^{(m)}$ of simple submodules ${}_R N_j^{(m)}$. As ${}_R M = \sum_{m \in M} Rm$, it follows that ${}_R M$ is the sum of simple submodules and therefore, by Proposition 6.1.8, is semisimple. □

We can now introduce the class of rings this book is centred around.

Definition 6.1.13 A unital ring R satisfying any, and hence every, condition in Theorem 6.1.12 is called *(classically) semisimple* or *completely reducible*.

Remarks 6.1.14

1. Strictly speaking we defined 'left semisimple ring' as we are working in R-$\mathcal{M}od$. An analogous concept of a 'right semisimple ring' in $\mathcal{M}od$-R turns out to be an equivalent notion (see Corollary 7.1.7) wherefore we drop the specification right from the start.
2. Theorem 6.1.12 (e) states: R is semisimple if and only if $_R R$ is semisimple. The submodules of $_R R$ are the left ideals of R and the simple submodules are the minimal left ideals. We thus see that a semisimple ring can be reconstructed from its 'smallest building blocks'.
3. We emphasise "classically" in the above definition since there is another, more modern version of "semisimple ring" which was introduced by N. Jacobson. This concept, which we will briefly discuss in Chap. 7 and in some special cases in Chap. 10, is also referred to as "semiprimitive" or "J-semisimple" and avoids inherent finiteness conditions. Therefore it is more suited for applications in infinite dimensions that are typical in Functional Analysis, for example.

6.2   Projective and Injective Modules

In this section we continue our study of the nice properties that semisimple rings enjoy; this will lead us to the Artin–Wedderburn theorem in the next chapter. To this end, we introduce two extremely important classes of modules.

Definition 6.2.1 Let R be a ring. The module $_R P \in R$-$\mathcal{M}od$ is said to be *projective* if it satisfies the following condition:

Whenever two modules $_R M, _R N \in R$-$\mathcal{M}od$ and an epimorphism $f \in \mathrm{Hom}_R ({}_R M, {}_R N)$ as well as a homomorphism $g_0 \in \mathrm{Hom}_R ({}_R P, {}_R N)$ are given, there exists $g \in \mathrm{Hom}_R ({}_R P, {}_R M)$ such that $fg = g_0$.
("Every homomorphism from $_R P$ can be lifted.")

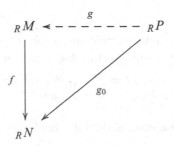

Definition 6.2.2 Let R be a ring. The module $_RI \in R\text{-}\mathcal{M}od$ is said to be *injective* if it satisfies the following condition:

Whenever two modules $_RM, {}_RN \in R\text{-}\mathcal{M}od$ and a monomorphism $f \in \text{Hom}_R ({}_RN, {}_RM)$ as well as a homomorphism $g_0 \in \text{Hom}_R ({}_RN, {}_RI)$ are given, there exists $g \in \text{Hom}_R ({}_RM, {}_RI)$ such that $gf = g_0$.

("Every homomorphism into $_RI$ can be extended.")

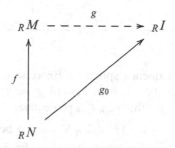

These two concepts from Homological Algebra allow us to formulate the following characterisation of semisimple rings.

Theorem 6.2.3 *The following conditions on a unital ring R are equivalent.*

(a) *R is semisimple;*
(b) *every left R-module is projective;*
(c) *every left R-module is injective.*

Proof (a) $\Rightarrow$ (b) Let $_RP \in R\text{-}\mathcal{M}od$ be part of the following diagram in which f is an epimorphism

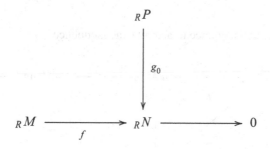

We can extend the exact sequence to a short exact sequence as follows

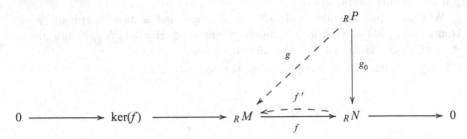

By Theorem 6.1.12, this sequence splits; by Exercise 6.4.4, there is thus $f' \in$ $\mathrm{Hom}_R\,(_RN,\,_RM)$ such that $ff' = \mathrm{id}_N$. Putting $g = f'g_0 \in \mathrm{Hom}_R\,(_RP,\,_RM)$ we find that $fg = ff'g_0 = g_0$; that is, $_RP$ is projective.

(b) $\Rightarrow$ (a) Let $0 \longrightarrow {}_RK \longrightarrow {}_RM \xrightarrow{f} {}_RN \longrightarrow 0$ be a short exact sequence in $R\text{-}\mathcal{M}od$. By hypothesis, $_RN$ is projective; thus, for $g_0 = \mathrm{id}_N$, we find $f' \in$ $\mathrm{Hom}_R\,(_RN,\,_RM)$ with $ff' = g_0 = \mathrm{id}_N$, that is, f' is a section for f. It follows that our sequence splits, and Theorem 6.1.12 yields that R is semisimple.

(a) $\Rightarrow$ (c) Let $_RI \in R\text{-}\mathcal{M}od$ be part of the following diagram in which f is a monomorphism

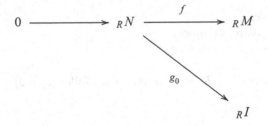

We complete the exact sequence to a short exact sequence

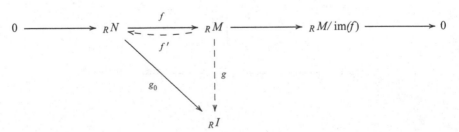

By Theorem 6.1.12, this sequence splits; by Exercise 6.4.4, there is thus $f' \in$ $\mathrm{Hom}_R\,(_RM,\,_RN)$ such that $f'f = \mathrm{id}_N$. Putting $g = g_0 f' \in \mathrm{Hom}_R\,(_RM,\,_RI)$ we find that $gf = g_0 f'f = g_0$ as desired; as a result, $_RI$ is injective.

(c) $\Rightarrow$ (a) Let $0 \longrightarrow {}_RK \xrightarrow{f} {}_RM \longrightarrow {}_RN \longrightarrow 0$ be a short exact sequence in $R\text{-}\mathcal{M}od$. By hypothesis, ${}_RK$ is injective; thus, for $g_0 = \mathrm{id}_K$, we find $f' \in \mathrm{Hom}_R({}_RM, {}_RK)$ with $f'f = g_0 = \mathrm{id}_K$. Therefore the sequence splits, and Theorem 6.1.12 entails that R is semisimple. $\qquad\square$

A major step forward in the direction of the Artin–Wedderburn theorem is provided by the next result revealing various finiteness properties of semisimple modules.

Theorem 6.2.4 *Let R be a unital ring. The following conditions on a semisimple module ${}_RM \in R\text{-}\mathcal{M}od$ are equivalent.*

(a) *${}_RM$ is finitely generated;*
(b) *${}_RM$ is finitely cogenerated;*
(c) *${}_RM$ is Noetherian;*
(d) *${}_RM$ is Artinian;*
(e) *${}_RM$ is a direct sum of finitely many simple submodules.*

Proof The proof is organised in the following way

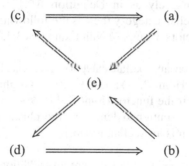

The implications (c) $\Rightarrow$ (a) and (d) $\Rightarrow$ (b) follow from Theorems 4.1 and 5.1.3, respectively. In order to show (a) $\Rightarrow$ (e), suppose ${}_RM = \bigoplus_{i \in I} {}_RM_i$ is a decomposition of ${}_RM$ into a direct sum of simple submodules ${}_RM_i$. As ${}_RM$ is finitely generated, there is a finite subset $J \subseteq I$ such that $\left(\bigcup_{j \in J} {}_RM_j\right) = \bigoplus_{j \in J} {}_RM_j = {}_RM$, by Exercise 4.3.7. Since ${}_RM_i \cap {}_RM_j = 0$ for $i \neq j$, it follows that $J = I$ so that (e) holds.

In order to show (b) $\Rightarrow$ (e) suppose ${}_RM$ is finitely cogenerated. Let ${}_RM = \bigoplus_{i \in I} {}_RM_i$ be a decomposition of ${}_RM$ into a direct sum of simple submodules ${}_RM_i$. Fix $j \in J$ and put ${}_RN_j = \bigoplus_{i \neq j} {}_RM_i$. In this way, we obtain a family of submodules of ${}_RM$ with the property $\bigcap_{j \in I} {}_RN_j = 0$, as ${}_RM_j \not\subseteq {}_RN_j$. By Exercise 5.5.1, there is a finite subset $J \subseteq I$ such that $\bigcap_{j \in J} {}_RN_j = 0$. Let $k \in I \setminus J$; then ${}_RM_k \not\subseteq {}_RM_j$ for any $j \in J$ by simplicity. Hence ${}_RM_k \subseteq {}_RN_j$ for all $j \in J$. It follows that ${}_RM_k = 0$. As a result, $I \setminus J = \emptyset$, that is, $I = J$ is finite and (e) holds.

Finally suppose that $_R M = \bigoplus_{i \in I} {_R M_i}$ for a finite set I. If all $_R M_i$ are simple, they are both Noetherian as well as Artinian and $_R M$ inherits these properties by Corollaries 4.1.3 and 5.1.6, respectively. This shows (e) $\Rightarrow$ (c) and (e) $\Rightarrow$ (d). $\square$

From this theorem, we gain a lot of insight into the structure of a semisimple ring R as the module $_R R$ is clearly finitely generated.

Corollary 6.2.5 *Every semisimple unital ring R is both left Noetherian as well as left Artinian. Moreover, $R = \bigoplus_{i=1}^{n} L_i$ for a finite set $\{L_1, \ldots, L_n\}$ of minimal left ideals of R.*

It is clear that there is a right-handed version for each of the last few results. However, we aim for a symmetric version of the above corollary which will be achieved in the next chapter.

6.3 Projective and Injective Objects

In a general category, objects which behave similarly to projective and injective modules in R-$\mathcal{M}od$ play an important role, especially in Homological Algebra. They can be defined precisely as in Definition 6.2.1 and in Definition 6.2.2, respectively, but in an abelian category there is the following equivalent description in terms of the Hom-functors. Compare with Examples 3.2.2 and Sect. 4.2.3.

Definition 6.3.1 Let $\mathscr{C}$ be an abelian category. An object $P \in ob(\mathscr{C})$ is called *projective* if the functor $Hom(P, -) : \mathscr{C} \to \mathscr{A}\mathscr{G}r$ is right exact. An object $I \in ob(\mathscr{C})$ is called *injective* if the functor $Hom(-, I) : \mathscr{C} \to \mathscr{A}\mathscr{G}r$ is right exact. (We followed here the usual convention to denote the morphism set $Mor_{\mathscr{C}}(A, B)$, $A, B \in ob(\mathscr{C})$ by $Hom(A, B)$ as it is an abelian group.)

Sometimes, to be injective does not require any additional property; e.g., in the category of vector spaces, every object is injective. On the other hand, in $\mathscr{A}\mathscr{G}r$, the injective objects are precisely the divisible abelian groups and the full subcategory of finite abelian groups has no projective objects at all. We shall focus on injective objects in the following, trying to illustrate what they might be good for. Evidently, projectivity is the dual concept to injectivity.

In some categories the requirement on an injective object according to the above definition is too restrictive: e.g., in $\mathscr{B}an_\infty$, the category of all Banach spaces, the categorical definition would imply that $\mathbb{C}$ is not injective. To see this, let ℓ^1 denote the Banach space of all absolutely summable sequences with the norm $\|(\xi_n)_{n \in \mathbb{N}}\|_1 = \sum_{n=1}^{\infty} |\xi_n|$ and let c_0 denote the Banach space of all null sequences (sequences converging to 0) with the norm $\|(\xi_n)_{n \in \mathbb{N}}\|_\infty = \sup_n |\xi_n|$. Define a contractive linear functional $g_0 : \ell^1 \to \mathbb{C}$ by $g_0((\xi_n)_{n \in \mathbb{N}}) = \sum_{n=1}^{\infty} \xi_n$. Upon embedding ℓ^1 into c_0 canonically we see that g_0 cannot be extended to a bounded linear functional $g : c_0 \to \mathbb{C}$. Since $\mathbb{C}$ is complemented in every non-zero Banach space, it follows that 0 is the only injective object according to the categorical definition. (Use Exercise 6.4.5 to see this.) An analogous argument shows that 0 is the only projective one, too.

The solution to this problem is to restrict the class of admissible monomorphisms.

Definition 6.3.2 Let M be a class of morphisms in the category $\mathscr{C}$ which consists of monomorphisms in $\mathscr{C}$ and contains all isomorphisms. Assume further that M is closed under composition. The object $I \in \mathrm{ob}(\mathscr{C})$ is said to be M-*injective* if it satisfies the following condition:

Whenever two objects $A, B \in \mathrm{ob}(\mathscr{C})$ and a monomorphism $f \in M$, $f : A \to B$ as well as a morphism $g_0 \in \mathrm{Mor}_{\mathscr{C}}(A, I)$ are given, there exists $g \in \mathrm{Mor}_{\mathscr{C}}(B, I)$ such that $gf = g_0$.

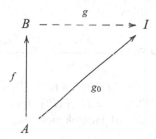

The monomorphisms in M are often referred to as *embeddings*; e.g., the good embeddings in the category of Banach spaces are the isometric linear mappings. (Note, however, that with this class M of embeddings, M-injectivity in $\mathscr{B}an_\infty$ is still different from M-injectivity in $\mathscr{B}an_1$: only the latter one gives the "extension under preservation of the norm" property which is the preferred one in Functional Analysis (Hahn–Banach theorem).)

Suppose M is a class of embeddings in the category $\mathscr{C}$. We say $\mathscr{C}$ *has enough* M-*injectives* if for every object $A \in \mathrm{ob}(\mathscr{C})$ there are an M-injective object I and an embedding $\iota \in \mathrm{Mor}_{\mathscr{C}}(A, I)$. For instance, $\mathscr{B}an_1$ has enough M-injectives, where M is the class of all linear isometries. This is a consequence of the Hahn–Banach theorem which allows to embed every Banach space into $\ell^\infty(\Lambda)$ for a suitable index set Λ (use Exercise 6.4.12 which also holds for non-abelian categories to show that $\ell^\infty(\Lambda)$ is M-injective). For a unital ring R, the category $R\text{-}\mathscr{M}od$ has enough M-injectives (where M is simply the class of all monomorphisms). This follows from the fact that $\mathrm{Hom}_{\mathbb{Z}}(R, G)$ is injective in $R\text{-}\mathscr{M}od$ for every divisible abelian group. For the details, see [29], Sect. 2.4 or [21], Sect. I.8.

Let $A \in \mathrm{ob}(\mathscr{C})$ be an object in the abelian category $\mathscr{C}$ with a fixed class M of embeddings. An *injective resolution of* A consists of a family $\{I^n \mid n \in \mathbb{N}_0\}$ of M-injective objects in $\mathscr{C}$ together with a family of morphisms $\{d^{(n)} \mid n \in \mathbb{N}_0\}$, where $d^{(n)} \in \mathrm{Mor}_{\mathscr{C}}(I^n, I^{n+1})$, as well as an M-morphism $A \to I^0$ such that the sequence

$$0 \xrightarrow{\hspace{2cm}} A \xrightarrow{\hspace{2cm}} I^0 \xrightarrow{d^{(0)}} I^1 \xrightarrow{d^{(1)}} I^2 \xrightarrow{d^{(2)}} \cdots$$

is exact. It can be shown that, if $\mathscr{C}$ has enough M-injectives, then every object in $\mathscr{C}$ has an injective resolution.

Let $F \colon \mathscr{C} \to \mathscr{A}\mathscr{G}r$ be a left exact functor (compare Sect. 4.2.3), where $\mathscr{C}$ is an abelian category with enough M-injectives. The *right derived functors of* F, R^nF, $n \in \mathbb{N}_0$, are defined as follows. For an object A in $\mathscr{C}$, choose an injective resolution as above and consider the complex

$$0 \xrightarrow{\hspace{2cm}} F(A) \xrightarrow{\hspace{2cm}} F(I^0) \xrightarrow{F(d^{(0)})} F(I^1) \xrightarrow{F(d^{(1)})} F(I^2) \xrightarrow{F(d^{(2)})} \cdots$$

and denote by $H^n(F(I^n)) = \ker F(d^{(n)})/\mathrm{im}F(d^{(n-1)})$, $n \geq 1$ the *nth cohomology group* of the complex. Then $R^nF(A) = H^n(F(I^n))$ for $n \geq 1$ and $R^0F(A) = F(A)$. Derived functors are at the heart of Homological Algebra; for more information, see, e.g., [21] or [29].

6.4 Exercises

Exercise 6.4.1 Let R be a unital ring, and let $_RM \in R\text{-}\mathscr{M}\!\mathit{od}$. Show that the following conditions on $_RM$ are equivalent.

(a) $_RM$ is simple;
(b) $_RM$ is cyclic and every non-zero element $m \in {}_RM$ generates $_RM$;
(c) $_RM \cong R/I$ for some maximal left ideal I of R.

Exercise 6.4.2 (The Original Formulation of Schur's Lemma) Let G be a group and let $\rho \colon G \to GL_n(\mathbb{C})$ be a finite-dimensional irreducible representation of G. Put $V = \mathbb{C}^n$ and let $R = \mathbb{C}[G]$. Show that the only endomorphisms of $_RV$ are the multiplications by elements of $\mathbb{C}$. (You may use that the only finite-dimensional division algebra over $\mathbb{C}$ is $\mathbb{C}$ itself.)

Exercise 6.4.3 Show that the index sets of any two decompositions of a non-zero semisimple module into a direct sum of simple submodules have the same cardinality.

Exercise 6.4.4 Let R be a unital ring. Let

$$0 \xrightarrow{} {}_R M_1 \xrightarrow{f} {}_R M_2 \xrightarrow{g} {}_R M_3 \xrightarrow{} 0$$

be a short exact sequence in $R\text{-}\mathscr{Mod}$. Show that the sequence splits if and only if either of the following conditions holds.

(a) $\exists\, f' \in \mathrm{Hom}_R\,({}_R M_2,\, {}_R M_1)$: $\quad f'f = \mathrm{id}_{M_1}$ (that is, f is a *section*);
(b) $\exists\, g' \in \mathrm{Hom}_R\,({}_R M_3,\, {}_R M_2)$: $\quad gg' = \mathrm{id}_{M_3}$ (that is, g is a *retraction*).

With this notation, the following decomposition holds

$$ {}_R M_2 = \mathrm{im}(f) \oplus \ker(f') = \ker(g) \oplus \mathrm{im}(g'). $$

Exercise 6.4.5 In generalising the previous exercise to an abelian category $\mathscr{C}$, we say that an object $E \in \mathrm{ob}(\mathscr{C})$ is a *retract of* $F \in \mathrm{ob}(\mathscr{C})$ if there exist morphisms $s \in \mathrm{Hom}(E, F)$ and $r \in \mathrm{Hom}(F, E)$ such that $rs = 1_E$. In this case we call s a *section* and r a *retraction*. An object $E \in \mathrm{ob}(\mathscr{C})$ is an *absolute* M-*retract* if every $\iota\colon E \to F$ in M for any $F \in \mathrm{ob}(\mathscr{C})$ is a section (where M is a class of embeddings as in the text above).

Show that every M-injective object is an absolute M-retract. Every retract of an M-injective object is M-injective. If $\mathscr{C}$ has enough M-injectives then every absolute M-retract is M-injective.

Exercise 6.4.6 Apply the characterisation of semisimple modules to show that every vector space over a field has a basis.

Exercise 6.4.7 Let ${}_R P \in R\text{-}\mathscr{Mod}$ for a unital ring R. Show that the following conditions on ${}_R P$ are equivalent.

(a) ${}_R P$ is projective;
(b) ${}_R P$ is (isomorphic to) a direct summand of a free left R-module;
(c) every epimorphism onto ${}_R P$ splits.

Exercise 6.4.8 The covariant Hom-functor $\mathrm{Hom}_R\,({}_R U, -)\colon R\text{-}\mathscr{Mod} \to \mathscr{AGr}$ is left exact for every ${}_R U \in R\text{-}\mathscr{Mod}$, see Example 4.2.3.1 in Chap. 4. Show that $\mathrm{Hom}_R\,({}_R U, -)$ is right exact if and only if ${}_R U$ is projective.

Exercise 6.4.9 Show that a unital ring R which is simple as a left R-module is a division ring.

Exercise 6.4.10 Let $_R M \in R\text{-}\mathcal{M}od$ for a unital ring R and let $n \in \mathbb{N}$. Given $f \in \text{End}_R (_R M^n)$, let $\alpha_{k\ell}$ be the composition

$$0 \longrightarrow {}_R M \xrightarrow{\;i_\ell\;} {}_R M^n \xrightarrow{\;f\;} {}_R M^n \xrightarrow{\;p_k\;} {}_R M \longrightarrow 0$$

of the projection p_k onto the kth factor with f with the injection i_ℓ of the ℓth summand. Show that $f \mapsto (\alpha_{k\ell})$ defines a ring isomorphism between $\text{End}_R (_R M^n)$ and $M_n (\text{End}_R (_R M))$.

Exercise 6.4.11 Show that, for every unital ring R and each $n \in \mathbb{N}$, the centre $Z(M_n(R))$ of $M_n(R)$ is isomorphic to $Z(R)$.

Exercise 6.4.12 Suppose $\mathscr{C}$ is an (abelian) category which is bicomplete, that is, arbitrary products and coproducts exist. Show that, if $\{P_\alpha\}$ is a family of projective objects in $\mathscr{C}$, then $\coprod_\alpha P_\alpha$ is projective and that, if $\{I_\alpha\}$ is a family of injective objects in $\mathscr{C}$, then $\prod_\alpha I_\alpha$ is injective.

Exercise 6.4.13 Let $n \in \mathbb{N}$. Show that

$$0 \longrightarrow \mathbb{Z}_n \longrightarrow \mathbb{Q}/\mathbb{Z} \longrightarrow \mathbb{Q}/\mathbb{Z} \longrightarrow 0,$$

with the canonical mappings and $\mathbb{Z}_n = \mathbb{Z}/n\mathbb{Z}$, is an injective resolution of $\mathbb{Z}_n$ but $\mathbb{Z}_n$ itself is not injective (as a $\mathbb{Z}$-module).

The Artin–Wedderburn Theorem

7

In this chapter, we shall obtain the full structure theorem for semisimple rings. This result, due to J. H. M. Wedderburn (1907) for semisimple finite-dimensional algebras and to E. Artin (1927) in the general case, enables us to determine completely this class of rings from the more elementary class of division rings. It is generally regarded as the first major result in the abstract structure theory of rings. In Sect. 7.2 below, we will briefly discuss the semisimplicity of group rings via Maschke's Theorem 7.2.1. A detailed proof of the Hopkins–Levitzki theorem, already stated in Theorem 5.3.2, is given in Sect. 7.3.

7.1 The Structure of Semisimple Rings

We start by recalling a well-known proposition from Ring Theory.

Proposition 7.1.1 *Let R be a unital ring, and let $n \in \mathbb{N}$. Every ideal J in $M_n(R)$ is of the form $J = M_n(I)$ for some ideal I in R. Thus, $M_n(R)$ is simple if and only if R is simple.*

Proof Evidently, if $I \subseteq R$ is an ideal, then $M_n(I)$ is an ideal of $M_n(R)$. Conversely, suppose $J \subseteq M_n(R)$ is an ideal. Let e_{ij}, $1 \leq i, j \leq n$ denote the canonical matrix units, that is, the entries of e_{ij} are all 0 but for the ith row and jth column where the entry is 1. Define $I \subseteq R$ as the set of all $a \in R$ such that $ae_{11} \in J$. Clearly I is an ideal in R. We aim to show that $J = M_n(I)$.

Let $a = (a_{ij}) \in M_n(R)$. For all i, j, k, ℓ, we have $e_{ij}ae_{k\ell} = a_{jk}e_{i\ell}$. Thus, if $a \in J$, choosing $i = \ell = 1$, we find that $a_{jk}e_{11} \in J$ and therefore $a_{jk} \in I$ for all j, k. Consequently $J \subseteq M_n(I)$. Conversely, let $a = (a_{ij}) \in M_n(I)$. By definition of I, $a_{ij}e_{11} \in J$ for all i, j. It follows that $a_{ij}e_{ij} = e_{i1}(a_{ij}e_{11})e_{1j} \in J$ and thus, $a = \sum_{i,j} a_{ij}e_{ij} \in J$. Hence $M_n(I) \subseteq J$.

The final statement is now clear. □

© The Author(s), under exclusive license to Springer Nature Switzerland AG 2022
M. Mathieu, *Classically Semisimple Rings*,
https://doi.org/10.1007/978-3-031-14209-3_7

In the next result we collect a number of properties of matrix rings over division rings.

Theorem 7.1.2 *Let D be a division ring, and let $R = M_n(D)$ for some $n \in \mathbb{N}$. Then*

(i) *R is simple, semisimple, left Artinian and left Noetherian;*
(ii) *up to isomorphism, R has a unique simple module $_RV \in R\text{-}\mathcal{M}od$. On it, R acts faithfully and $_RR \cong {_R}V^n$;*
(iii) *$\mathrm{End}_R\,(_RV) \cong D$.*

Proof Every division ring is simple; hence R is simple by Proposition 7.1.1. Let V be the n-tuple column space D^n, viewed as a right D-vector space. Then $R = M_n(D)$ acts on the left of V by matrix multiplication, so $_RV \in R\text{-}\mathcal{M}od$ is a faithful unital left R-module. In fact, $M_n(D)$ can be identified with $\mathrm{End}_D(V_D)$ in the usual way by choosing a basis of V_D. An argument analogous to the one in Examples 6.1.3 shows that $_RV$ is a simple R-module.

For each $i \in \{1, \ldots, n\}$, let L_i denote the left ideal of R consisting of those $n \times n$-matrices over D whose columns other than the ith are zero. Then $R = L_1 \oplus \ldots \oplus L_n$ and, since $_RL_i \cong {_R}V$, we get $_RR \cong {_R}V^n$. By Proposition 6.1.8, $_RR$ is semisimple so the ring R is (left) semisimple. By Corollary 6.2.5, R is thus left Artinian as well as left Noetherian, which completes the proof of (i).

The uniqueness of $_RV$ stated in (ii) now follows from Proposition 6.1.9.

In order to prove (iii), we define a mapping $\sigma : D \to \mathrm{End}_R\,(_RV)$ by $v \cdot \sigma(d) = v \cdot d$, $v \in V$, $d \in D$. Note that we write here endomorphisms on the right in order to avoid the opposite ring D^{op}; see Remark 7.1.3 below. Clearly, σ is a unital ring homomorphism and, since D acts faithfully on V, σ is injective. To show that σ is surjective, let $f \in \mathrm{End}_R\,(_RV)$. Then $\begin{pmatrix}1\\0\\\vdots\\0\end{pmatrix} f = \begin{pmatrix}d\\ *\\\vdots\\ *\end{pmatrix}$ for some $d \in D$ and hence,

$$\begin{pmatrix}a_1\\\vdots\\a_n\end{pmatrix} f = \left(\begin{pmatrix}a_1 & 0 & \ldots & 0\\\vdots & & \mathbf{0}\\a_n & & \end{pmatrix}\begin{pmatrix}1\\0\\\vdots\\0\end{pmatrix}\right) f = \begin{pmatrix}a_1 & \ldots & \\\vdots & \mathbf{0}\\a_n & & \end{pmatrix}\left(\begin{pmatrix}1\\0\\\vdots\\0\end{pmatrix} f\right)$$

$$= \begin{pmatrix}a_1 & \ldots & \\\vdots & \mathbf{0}\\a_n & & \end{pmatrix}\begin{pmatrix}d\\ *\\\vdots\\ *\end{pmatrix} = \begin{pmatrix}a_1 d\\\vdots\\a_n d\end{pmatrix} = \begin{pmatrix}a_1\\\vdots\\a_n\end{pmatrix}\sigma(d)$$

so that $f = \sigma(d)$.   □

Remarks 7.1.3

1. By Schur's Lemma (6.1.4), we know that $\mathrm{End}_R\,(_R V)$ is a division ring as $_R V$ is simple. The point in the above theorem is that this division ring is the one we start with.
2. The notational change in the proof of Theorem 7.1.2 to write endomorphisms on the right allows us to avoid the opposite ring D^{op}. It occurs naturally here since right multiplication $\rho_r : x \mapsto xr$ is a left R-module map on R and the composition $\rho_s \rho_r$ of ρ_r and ρ_s is ρ_{rs} and not ρ_{sr} so that the map $r \mapsto \rho_r$ is not a ring homomorphism.

Example 7.1.4 Let $D_1, \ldots, D_\ell$ be division rings and let $n_1, \ldots, n_\ell \in \mathbb{N}$. Then $R = M_{n_1}(D_1) \times \ldots \times M_{n_\ell}(D_\ell)$ is a unital semisimple ring. This follows from Theorem 7.1.2 together with the fact that every finite direct product of semisimple rings is again semisimple.

Historical Note Joseph Henry Maclagan Wedderburn was born in 1882 in Scotland as the 10th of 14 children. He entered the University of Edinburgh at the age of 16 and was elected Fellow of the Royal Society of Edinburgh when he was only 21 years old. After further studies at Leipzig, Berlin and Chicago he returned to Edinburgh in 1905 and was awarded his DSc and his PhD in 1908. Wedderburn served in the First World War between 1914 and 1918 after which he became an Associate Professor at Princeton in 1921. Among his (three) PhD students at Princeton was Nathan Jacobson. He made important advances in the theory of rings, algebras and matrix theory; among these is his celebrated result that every finite division ring is a field which also has important consequences for finite projective geometry. His 1907 paper "On hypercomplex numbers" contains the structure theorem for finite-dimensional semisimple algebras. See also Artin's paper [5]. Wedderburn died in Princeton in October 1948.
https://www-history.mcs.st-andrews.ac.uk/Biographies/Wedderburn.html

We shall now see that the above Example 7.1.4 already exhausts all possibilities.

7.1.5 Artin–Wedderburn Theorem *Every semisimple unital ring is isomorphic to a finite direct product of matrix rings over division rings.*

Proof Let R be a semisimple unital ring. By Corollary 6.2.5, there are finitely many minimal left ideals $L_1, \ldots, L_n$ in R such that

$$R = L_1 \oplus \ldots \oplus L_n.$$

Some of the L_i's may be isomorphic to each other as left R-modules; from each isomorphism class we pick one representative and denote this simple submodule of $_RR$ by $_RM_j$ (so that $_RM_j \not\cong {}_RM_i$ for $j \neq i$). Suppose we have ℓ such isomorphism classes. Then there exist $n_1, \ldots, n_\ell \in \mathbb{N}$ such that

$$_RR \cong {}_RM_1^{n_1} \oplus \ldots \oplus {}_RM_\ell^{n_\ell}. \tag{7.1.1}$$

The $_RM_i^{n_i}$ are called the *homogeneous components* of R. By Schur's Lemma (6.1.4), $\mathrm{End}_R\,(_RM_i) = D_i$ is a division ring for each $1 \leq i \leq \ell$. Note that $\mathrm{End}_R\,(_RM_i^{n_i}) \cong M_{n_i}(\mathrm{End}_R\,(_RM_i)) = M_{n_i}(D_i)$ for each $1 \leq i \leq \ell$ by Exercise 6.4.10.

From identity (7.1.1) we obtain

$$\mathrm{End}_R\,(_RR) \cong \mathrm{End}_R\Big(\bigoplus_{i=1}^{\ell} {}_RM_i^{n_i}\Big)$$

$$= \mathrm{Hom}_R\Big(\bigoplus_{i=1}^{\ell} {}_RM_i^{n_i}, \bigoplus_{j=1}^{\ell} {}_RM_j^{n_j}\Big)$$

$$\cong \prod_{i=1}^{\ell} \mathrm{Hom}_R\Big(_RM_i^{n_i}, \bigoplus_{j=1}^{\ell} {}_RM_j^{n_j}\Big)$$

$$= \prod_{i=1}^{\ell} \mathrm{Hom}_R\Big(_RM_i^{n_i}, \prod_{j=1}^{\ell} {}_RM_j^{n_j}\Big)$$

$$\cong \prod_{i=1}^{\ell} \prod_{j=1}^{\ell} \mathrm{Hom}_R\big(_RM_i^{n_i}, {}_RM_j^{n_j}\big),$$

where we used the universal properties of the direct sum and the direct product of modules and the fact that these constructions coincides with each other if the index set is finite, see Exercise 2.7.10.

We now apply a similar chain of arguments to compute the individual homomorphism groups.

$$\mathrm{Hom}_R\big(_RM_i^{n_i}, {}_RM_j^{n_j}\big) = \mathrm{Hom}_R\Big(\bigoplus^{n_i} {}_RM_i, \prod^{n_j} {}_RM_j\Big)$$

$$\cong \prod^{n_i} \prod^{n_j} \mathrm{Hom}_R\big(_RM_i, {}_RM_j\big)$$

$$= \begin{cases} \displaystyle\prod^{n_i} \prod^{n_i} \mathrm{Hom}_R\big(_RM_i, {}_RM_i\big) & \text{if } i = j \\ 0 & \text{if } i \neq j, \end{cases}$$

where we used that $\mathrm{Hom}_R\left({}_R M_i, {}_R M_j\right) = 0$ for $i \neq j$ since both modules are simple and non-isomorphic. Putting these two computations together we find

$$\mathrm{End}_R\left({}_R R\right) \cong \prod_{i=1}^{\ell} \mathrm{Hom}_R\left({}_R M_i^{n_i}, {}_R M_i^{n_i}\right)$$

$$= \prod_{i=1}^{\ell} \mathrm{End}_R\left({}_R M_i^{n_i}\right)$$

$$\cong \prod_{i=1}^{\ell} M_{n_i}(D_i).$$

Finally observe that $R \cong \mathrm{End}_R\left({}_R R\right)$ via $r \mapsto \rho_r$ if we write the endomorphisms on the right, compare Exercise 2.7.1. Consequently we arrive at

$$R \cong M_{n_1}(D_1) \times \ldots \times M_{n_\ell}(D_\ell)$$

which was our aim. $\qquad \square$

A representation as in the above result is unique in various ways, as we shall see below in Theorem 7.1.9 and in Exercise 7.4.1.

Remark 7.1.6 If we want to avoid writing endomorphisms on the right in the above proof, we obtain instead

$$R \cong \mathrm{End}_R\left({}_R R\right)^{\mathrm{op}} \cong \prod_{i=1}^{\ell} M_{n_i}(D_i)^{\mathrm{op}} = \prod_{i=1}^{\ell} M_{n_i}(D_i^{\mathrm{op}}),$$

where the last equality is given by the transpose mapping.

Corollary 7.1.7 *A unital ring is left semisimple if and only if it is right semisimple.*

Proof This follows from the fact that we have an analogous theory of semisimple right modules and right semisimple rings leading to the same Artin–Wedderburn Theorem which is a symmetric statement. $\qquad \square$

Corollary 7.1.8 *A commutative unital ring is semisimple if and only if it is a finite direct product of fields.*

Proof While the "if"-part is immediate, the "only if"-part follows from Theorem 7.1.5 together with the fact that $M_n(D)$ is commutative if and only if $n = 1$ and D is commutative. $\qquad \square$

In addition, we have a very strong form of uniqueness in the decomposition in the Artin–Wedderburn Theorem.

7.1.9 Uniqueness Theorem for Semisimple Rings *Let R be a unital ring and suppose*

$$R = \prod_{i=1}^{n} R_i = \prod_{j=1}^{m} R'_j$$

are two decompositions with simple rings R_i, R'_j for $1 \leq i \leq n$, $1 \leq j \leq m$. Then $n = m$ and for every i there is a unique j such that $R_i = R'_j$.

Proof We will identify each R_i and R'_j with a two-sided ideal in R in the canonical way. Then $R_i = RR_i = R_iR$ and $R'_j = RR'_j = R'_jR$ for all i, j. Therefore

$$R_i = R_iR = \prod_{j=1}^{m} R_iR'_j \qquad (1 \leq i \leq n).$$

Since each $R_iR'_j$ is thus an ideal of R_i, and R_i is simple, we have $R_iR'_j = 0$ or $R_iR'_j = R_i$. As $R_i \neq 0$, there is some j such that $R_iR'_j \neq 0$ and thus $R_iR'_j = R_i$. On the other hand, $R_iR'_j = R'_j$ as R'_j is simple so that $R_i = R'_j$.

In the identification $R'_j \mapsto 0 \times \ldots \times R'_j \times 0 \ldots \times 0$, all R'_js are distinct from each other, as are the R_is. Therefore, for each R_i there can be only one R'_j with $R_i = R'_j$; this entails that $n = m$. $\square$

Putting Theorems 7.1.5 and 7.1.9 together we arrive at a very satisfactory result illustrating in a beautiful way what the structure theory of rings is about. This result is at the very heart of our book.

Corollary 7.1.10 *A unital ring R is semisimple if and only if it is the unique finite direct product of simple Artinian rings $R_1, \ldots, R_n$.*

Note that we must insert "Artinian" in the sufficiency condition as not every simple ring is semisimple; this will be illustrated in the example below.

Example 7.1.11 Let D be a division ring (for a concrete example, take $D = \mathbb{Q}$). For each $n \in \mathbb{N}$, let $R_n = M_{2^n}(D)$ which is a simple unital ring by Proposition 7.1.1. We define unital ring monomorphisms $R_n \to R_{n+1}$ by

$$M_{2^n}(D) \ni A \longmapsto \begin{pmatrix} A & 0 \\ 0 & A \end{pmatrix} \in M_{2^{n+1}}(D).$$

Identifying R_n with a unital subring of R_{n+1} in this way, we put $R = M_{2^\infty}(D) = \bigcup_{n=1}^{\infty} R_n$. (This is an instance of a "direct limit" of a directed system of rings.) Clearly, R is a unital ring.

In order to show that R is simple, let I be a non-zero ideal of R. Then there is some R_n such that $I \cap R_n \neq 0$. As $I \cap R_n$ is an ideal of R_n and R_n is simple, we obtain $I \cap R_n = R_n$, that is, $R_n \subseteq I$. Since $1_R = 1_{R_n} \in R_n$ it follows that $I = R$.

We next show that R is not left Artinian (and hence cannot be semisimple, by Corollary 6.2.5). For $n \in \mathbb{N}$, let e_n denote the matrix unit e_{11} in R_n. Then $e_{n+1} = e_{n+1}e_n \in R_{n+1}$. For instance,

$$e_1 \text{ in } R_1 = M_2(D) \quad \text{is} \quad \begin{pmatrix} 1 & 0 \\ 0 & 0 \end{pmatrix},$$

$$e_1 \text{ in } R_2 = M_4(D) \quad \text{is} \quad \begin{pmatrix} 1 & 0 & 0 & 0 \\ 0 & 0 & 0 & 0 \\ 0 & 0 & 1 & 0 \\ 0 & 0 & 0 & 0 \end{pmatrix},$$

$$e_2 \text{ in } R_2 = M_4(D) \quad \text{is} \quad \begin{pmatrix} 1 & 0 & 0 & 0 \\ 0 & 0 & 0 & 0 \\ 0 & 0 & 0 & 0 \\ 0 & 0 & 0 & 0 \end{pmatrix},$$

so $e_2 = e_2 e_1$. In this way, we obtain a descending chain of left ideals

$$Re_1 \supseteq Re_2 \supseteq \ldots \supseteq Re_n \supseteq Re_{n+1} \supseteq \ldots$$

Suppose $e_n \in Re_{n+1}$ for some n. Then $e_n = ae_{n+1}$ where $a \in R_j = M_{2^j}(D)$ for some $j > n$. The $(2^n + 1, 2^n + 1)$-entry of e_n in R_{n+1} is 1, however the $(2^n + 1, 2^n + 1)$-entry of ae_{n+1} is 0, which is impossible. As a result, the descending chain cannot become stationary and R is not left Artinian.

As we shall see now, the Artinian property is the one that turns a simple ring into a semisimple one.

Proposition 7.1.12 *For every simple unital ring R, the following conditions are equivalent.*

(a) *R is semisimple;*
(b) *R is left Artinian;*
(c) *$R = M_n(D)$ for some division ring D and $n \in \mathbb{N}$.*

Proof The implications (c) $\Rightarrow$ (a) and (c) $\Rightarrow$ (b) are provided by Theorem 7.1.2. (a) $\Rightarrow$ (b) is Corollary 6.2.5 and (a) $\Rightarrow$ (c) follows from Theorem 7.1.5.

(b) $\Rightarrow$ (a) If $_RR$ is Artinian it contains a simple submodule by Theorem 5.1.3, see also Exercise 7.4.2. Hence R contains a minimal left ideal. Define the socle $\mathrm{soc}(R)$ of R as the sum of all minimal left ideals (compare Exercise 7.4.3). As $\mathrm{soc}(R)$ is a non-zero ideal in the simple ring R, we obtain $R = \mathrm{soc}(R)$, that is, $_RR$ is a sum of simple submodules. By Proposition 6.1.8, $_RR$ is semisimple so that R is a semisimple ring.   □

This result shows that a right Artinian simple ring must be left Artinian (and vice versa) since a similar characterisation as in Proposition 7.1.12 is available. For this reason, one simply speaks of 'Artinian simple rings'.

Remark 7.1.13 For an approach to the Artin–Wedderburn theorem from the viewpoint of abelian categories and Morita equivalence the reader is invited to consult [30, Sect. 4.12].

7.2   Maschke's Theorem

In this short section we shall discuss an important class of semisimple rings that is related to group representations. In Sect. 1.1.5 we indicated the connections between finite-dimensional representations of a group G and modules over the group ring $K[G]$, where K is a field. Indeed, given a finite group G and a field K, there is a one-to-one correspondence between K-linear representations of G and finitely generated left $K[G]$-modules. It turns out that the relation between the order of G and the characteristic of K determines the structure of the group ring.

Theorem 7.2.1 (Maschke) *Let K be a field with characteristic q (finite or zero). Let G be a finite group such that q does not divide $|G|$. Then the group ring $K[G]$ is semisimple.*

Proof Set $R = K[G]$ and let the sequence $0 \longrightarrow {}_RN \xrightarrow{\ i\ } {}_RM$ in R-$\mathcal{M}od$ be exact. As N and M are K-vector spaces, we can define a linear splitting $p_0 \colon M \to N$ easily: simply extend a basis of N to a basis of M and define p_0 by $p_0(i(b)) = b$ for all basis elements in N and $p_0 = 0$ otherwise. Then $p_0 i = \mathrm{id}_N$. We intend to "upgrade" p_0 to a splitting map $p \in R$-$\mathcal{M}od$.

To this end, suppose that $|G| = n$ and define $p \colon M \to N$ by

$$p(m) = \frac{1}{n} \sum_{g \in G} g \cdot p_0(g^{-1}m) \qquad (m \in M).$$

For all $h \in G$, we have

$$p(hm) = \frac{1}{n} \sum_{g \in G} g \cdot p_0(g^{-1}hm) = \frac{1}{n} \sum_{g \in G} h(h^{-1}g) \cdot p_0((h^{-1}g)^{-1}m)$$

$$= h \cdot \frac{1}{n} \sum_{h^{-1}g \in G} (h^{-1}g) \cdot p_0((h^{-1}g)^{-1}m)$$

$$= h \cdot p(m)$$

as every element of G is uniquely expressed in the form $h^{-1}g$. Consequently, $p \in \operatorname{Hom}_R({}_RM, {}_RN)$. Since ${}_RN \leq {}_RM$, $g^{-1}m \in {}_RN$ for every $m \in {}_RN$; hence, for such m,

$$p(i(m)) = \frac{1}{n} \sum_{g \in G} g \cdot p_0(i(g^{-1}m)) = \frac{1}{n} \sum_{g \in G} g(g^{-1}m) = \frac{1}{n} \sum_{g \in G} m = m.$$

Thus $pi = \operatorname{id}_N$ and therefore ${}_RN$ is a direct summand of ${}_RM$. By Theorem 6.1.12, $K[G]$ is a semisimple ring. $\square$

On the other hand, it is not too hard to show that, if the characteristic of K does divide $|G|$, then $K[G]$ cannot be semisimple; see, e.g., [7, Proposition 4.1.5].

Since for every algebraically closed field K, the endomorphism ring $\operatorname{End}_{K[G]}({}_RM)$ of a simple left $K[G]$-module ${}_RM$ is isomorphic to K (Exercise 7.4.17), a combination of Theorems 7.1.5 and 7.2.1 yields the following result.

Corollary 7.2.2 *Let G be a finite group. Then the group ring $\mathbb{C}[G]$ is isomorphic to a finite direct sum of matrix rings over $\mathbb{C}$.*

A comprehensive discussion of the semisimplicity of group rings can be found in [24] and [31].

7.3 The Hopkins–Levitzki Theorem

We stated the Hopkins–Levitzki theorem (Theorem 5.3.2) already two chapters ago but found that a proof would be better placed after a thorough discussion of semisimple rings. Let us emphasise that *ring* will mean *unital ring* throughout.

We start with a generalisation of Definition 5.2.3 to the non-commutative setting.

Definition 7.3.1 For a ring R,

$$\text{rad}(R) = \bigcap \{L \mid L \text{ maximal left ideal in } R\}$$

is called the *Jacobson radical* of R.

Analogously, one could define a 'right-handed' version as the intersection of all maximal right ideals. However, this is unnecessary as the next result shows.

Proposition 7.3.2 *In every ring R, the following sets agree with each other:*

  (i) *the intersection of all maximal left ideals;*
 (ii) *the intersection of all maximal right ideals;*
(iii) $\{y \in R \mid 1 - xy \text{ is invertible for all } x \in R\};$
 (iv) $\{y \in R \mid 1 - yx \text{ is invertible for all } x \in R\};$
  (v) $\{y \in R \mid 1 - xyz \text{ is invertible for all } x, z \in R\}.$

In particular, the Jacobson radical $\text{rad}(R)$ *is an ideal of R.*

Proof We start with an observation which is either attributed to Jacobson or Kaplansky: for all $x, y \in R$,

$$1 - xy \text{ is invertible} \iff 1 - yx \text{ is invertible}.$$

By symmetry, we only have to verify one implication, and the observation will establish the equivalence of (iii) and (iv) in the proposition.
Suppose that $(1 - xy)u = 1 = u(1 - yx)$ for some $u \in R$. Then

$$(1 - yx)(yux + 1) = yux - yxyux + 1 - yx$$
$$= yux - y(u - 1)x + 1 - yx = 1,$$
$$(yux + 1)(1 - yx) = yux + 1 - yuxyx - yx$$
$$= yux + 1 - y(u - 1)x - yx = 1$$

show that $1 - yx$ is invertible with inverse $yux + 1$. Clearly (v) implies (iii), so assume (iii) holds. Then $1 - zxy$ is invertible for all $z, x \in R$ whence, by the observation, $1 - xyz$ is invertible, which was to show.
Suppose $y \in R$ is not in the intersection of all maximal left ideals of R and let L be a maximal left ideal such that $y \notin L$. Then $L + Ry = R$ and there is $x \in R$ with $1 - xy \in L$. It follows that $1 - xy$ cannot be (left) invertible as $L \neq R$ which shows that the set described in (iii) is contained in the intersection of all maximal left ideals. On the other hand, if $1 - xy$ is not invertible for some $x \in R$ then the left

ideal $R(1 - xy)$ is proper, thus contained in a maximal left ideal L (by a standard application of Zorn's lemma). It follows that $y \notin L$ which establishes the reverse inclusion.

The equality of the sets in (ii) and (iv) is shown analogously. □

Corollary 7.3.3 *For every ring R, we have*

$$\text{rad}(R) = \bigcap \{\text{Ann}_R(M) \mid {}_R M \in R\text{-}\mathcal{M}od \text{ simple}\}.$$

Proof Let ${}_R M \in R\text{-}\mathcal{M}od$ be simple and take $y \in R$. If $ym \neq 0$ for some $m \in M$ then $Rym = M$ and thus, there is $x \in R$ with $(1 - xy)m = 0$. As $m \neq 0$ it follows that $1 - xy$ is not invertible so $y \notin \text{rad}(R)$. Conversely, let L be a maximal left ideal of R. Then ${}_R M = R/L$ is a simple left R-module. If $y \subset R$ satisfies $yM = 0$ then $y \in L$ which proves "$\supseteq$" above. □

Definition 7.3.4 A ring R is called *semiprimitive* if $\text{rad}(R) = 0$, and it is called *left primitive* if it has a faithful simple left module.

By Corollary 7.3.3, every left primitive ring is semiprimitive, and by Proposition 7.3.2, 'semiprimitive' is left-right symmetric. Analogously, one can define a *right primitive* ring but it turns out that this, in general, is a concept different from left primitive; cf. [24, Sect. 11].

Remark 7.3.5 Semiprimitive rings are also called Jacobson semisimple, or J-semisimple for short. In Analysis, where one is mostly interested in infinite-dimensional algebras, which hardly satisfy any finiteness conditions on ideals, it has become customary to call semiprimitive algebras *semisimple* and the older version *classically semisimple*. See, e.g., [1], Sect. 5.3.

Corollary 7.3.6 *For every ring R, the quotient ring $R/\text{rad}(R)$ is semiprimitive.*

Proof Let us denote $R/\text{rad}(R)$ by $\hat{R}$ and its elements by $\hat{x}$ etc. Note that $\text{rad}(R)$ is a proper ideal as $1 \notin \text{rad}(R)$. Take $y \in \text{rad}(\hat{R})$; by Proposition 7.3.2, $1 - \hat{x}\hat{y}$ is invertible in $\hat{R}$ for all $\hat{x} \in \hat{R}$. Let $u \in R$ be such that $\hat{u}(1 - \hat{x}\hat{y}) = 1$ in $\hat{R}$. Then $u(1 - xy) = 1 + r$ for some $r \in \text{rad}(R)$. As $1 + r$ is invertible in R, by Proposition 7.3.2, $u(1 - xy)$ and hence $1 - xy$ is left invertible for all $x \in R$. A similar argument shows that $1 - xy$ is right invertible, thus invertible for all $x \in R$ which implies that $y \in \text{rad}(R)$. As a result, $\hat{y} = 0$. □

By Corollary 7.3.3, every simple left R-module is a simple left $\hat{R}$-module, where $\hat{R} = R/\text{rad}(R)$. Conversely, every simple left $\hat{R}$-module is a simple left R-module in a canonical way. (See Exercise 2.7.3.)

The reader may wonder why we introduced the Jacobson radical en route to a proof of the Hopkins–Levitzki theorem. The reason will be revealed to them now.

Proposition 7.3.7 *For every ring R, the following properties are equivalent.*

(a) *R is semisimple;*
(b) *R is left Artinian and semiprimitive.*

Proof (a) $\Rightarrow$ (b) Every semisimple ring is left Artinian, by Corollary 6.2.5, so we only need to show that $\text{rad}(R) = 0$. Since $\text{rad}(R)$ is a direct summand of $_R R$, there is a left ideal $L \subseteq R$ such that $L \oplus \text{rad}(R) = R$. Let $e \in L$, $f \in \text{rad}(R)$ be such that $e + f = 1$. Then $e^2 + fe = e$ and, as $fe \in \text{rad}(R) \cap L = 0$, it follows that $e^2 = e$. Similarly, $f^2 = f$ so that e and f are orthogonal idempotents in R. As $e = 1 - f$ is invertible, it follows that $e = 1$ and so $f = 0$. We conclude that $\text{rad}(R) = Rf = 0$ as claimed.

(b) $\Rightarrow$ (a) We first observe the following: let L be a non-zero left ideal of R. As R is left Artinian, L contains a minimal left ideal (use Theorem 5.1.3 (c)). Every minimal left ideal N of R is a direct summand of $_R R$. Indeed, since $\text{rad}(R) = 0$ and $N \neq 0$ there exists a maximal left ideal K of R not containing N. Then $N \cap K = 0$ and $_R R = N \oplus K$.

Now suppose that R is not semisimple. Take a minimal left ideal N_1 in R and write $_R R = N_1 \oplus K_1$ for a maximal left ideal K_1 as above. As $K_1 \neq 0$ there exists a minimal left ideal $N_2 \subseteq K_1$. As before, N_2 is a direct summand of $_R R$ and thus of K_1 (use Exercise 2.7.8). We therefore can write $K_1 = N_2 \oplus K_2$ for a left ideal $K_2 \subsetneq K_1$. By induction, we obtain a strictly descending chain of left ideals $K_1 \supsetneq K_2 \supsetneq K_3 \supsetneq \ldots$ which contradicts the assumption that R is left Artinian. $\square$

The fact that every semisimple ring is semiprimitive can also be deduced from the Artin–Wedderburn theorem together with Exercise 7.4.7 since simple rings are trivially semiprimitive. Of course, the argument above uses a less heavy tool.

We have now taken a substantial step towards the Hopkins–Levitzki theorem. If R is left Artinian with $\text{rad}(R) = 0$ then, by the above proposition, R is semisimple and hence left Noetherian, by Corollary 6.2.5. In particular, for any left Artinian ring R, $R/\text{rad}(R)$ is left Noetherian, by Corollaries 5.1.5 and 7.3.6. But, in the statement of Theorem 5.3.2, there is no assumption on the Jacobson radical so what seems to be an obstruction will have to be removed. We shall next see how an additional tool enables us to achieve this goal.

The idea is to use a short exact sequence of modules like

$$0 \longrightarrow \text{rad}(R) \longrightarrow {}_R R \longrightarrow R/\text{rad}(R) \longrightarrow 0 \qquad (7.3.1)$$

together with Theorem 4.1.1 to obtain that $_R R$ is Noetherian from the same properties of the outer modules. However, a priori, we do not know that $\text{rad}(R)$ is Noetherian, or even just finitely generated (though it will be once we have shown that R is left Noetherian). The remedy is to work with a suitable power of the radical instead.

Throughout the remainder of this section we will denote $\mathrm{rad}(R)$ *by* J *and* $R/\mathrm{rad}(R)$ *by* $\hat{R}$.

For some $n \in \mathbb{N}$, we replace (7.3.1) by

$$0 \longrightarrow J^{n-1} \longrightarrow R \longrightarrow R/J^{n-1} \longrightarrow 0 \qquad (7.3.2)$$

and use that, if R is left Artinian, then $\hat{R}$ is semisimple. The choice of n will become clear immediately.

Call an ideal I of R *nil* if every element in I is nilpotent and call I *nilpotent* if $I^n = 0$ for some $n \in \mathbb{N}$.

Lemma 7.3.8 *For every ring R, the Jacobson radical J contains each nil ideal of R. If R is left Artinian then J is nilpotent.*

Proof Let $z \in R$ be nilpotent, say, $z^k = 0$. Then $1 - z$ is invertible with inverse $1 + z + z^2 + \ldots + z^{k-1}$. Let I be a nil ideal in R and take $y \in I$. Then, for all $x \in R$, $xy \in I$ is nilpotent so $1 - xy$ is invertible. By Proposition 7.3.2, $y \in \mathrm{rad}(R)$ so $I \subseteq \mathrm{rad}(R)$.

Suppose R is left Artinian. The descending chain $(J^k)_{k \in \mathbb{N}}$ becomes stationary, say at n. Put $I = J^n$ and suppose that $I \neq 0$. As $I^2 = I$ the set

$$\{L \subseteq R \mid L \text{ is a left ideal and } IL \neq 0\}$$

is not empty, thus contains a minimal element, say L_0 (Theorem 5.1.3). Take $y \in L_0$ such that $Iy \neq 0$. Since $I\,Iy \neq 0$ and $Iy \subseteq L_0$ it follows that $Iy = L_0$. Therefore $y = xy$ for some $x \in I$ which implies that $(1 - x)y = 0$. Since $x \in J^n \subseteq J$, $1 - x$ is invertible so that $y = 0$, a contradiction. We conclude that $J^n = I = 0$. $\square$

The reader may want to compare the above proof with the argument in Proposition 5.2.5; see also Exercise 7.4.8. We observe that, since every nilpotent ideal is nil, the Jacobson radical of a left Artinian ring is the largest nil ideal. On the other hand, it is the smallest ideal such that the corresponding quotient is semisimple, as we shall observe now.

Proposition 7.3.9 *In a left Artinian ring R, the Jacobson radical is the smallest ideal of R whose corresponding quotient is semisimple.*

Proof By Corollary 7.3.6 and Proposition 7.3.7, R/J is semisimple. Conversely, let $K \subseteq R$ be an ideal such that R/K is a semisimple left R-module, and let $\rho: R \to R/K$ be the canonical epimorphism. As R/K is semisimple, there is a left ideal W in R, containing K, such that $\rho(J) \oplus \rho(W) = R/K$. Suppose that $W \neq R$. Then there is a maximal left ideal L of R which contains W. By definition, $J \subseteq L$ so

$J + W \subseteq L$. On the other hand, $J + W = R$ which is a contradiction. As a result, $W = R$ and $\rho(J) = 0$, in other words, $J \subseteq K$. $\square$

Note also that, if K is an ideal of the left Artinian ring R and $J \subseteq K$, then R/K is semisimple, as a homomorphic image of R/J.

We need another auxiliary result which can be obtained, for instance, using the convenient characterisation of the Jacobson radical contained in Proposition 7.3.2.

Lemma 7.3.10 *Let I be an ideal of the ring R such that $I \subseteq J$. Then $\mathrm{rad}(R/I) = J/I$.*

Proof Let $\pi : R \rightarrow R/I$ and $\sigma : R/I \rightarrow R/J$ be the canonical epimorphisms. Take $y \in \mathrm{rad}(R)$; then $1 - xy$ is invertible in R for all $x \in R$ and thus, $1 - \pi(x)\pi(y)$ is invertible in R/I for all $\pi(x)$. By Proposition 7.3.2, $\pi(y) \in \mathrm{rad}(R/I)$ so $\mathrm{rad}(R)/I \subseteq \mathrm{rad}(R/I)$.

Take $y \in R$ such that $\pi(y) \in \mathrm{rad}(R/I)$. Then $1 - \pi(x)\pi(y)$ is invertible in R/I and hence, $1 - \sigma(\pi(x))\sigma(\pi(y))$ is invertible in R/J. We have seen in the proof of Corollary 7.3.6 that this implies $y \in J$ and thus, $\pi(y) \in J/I$. This establishes the reverse inclusion $\mathrm{rad}(R/I) \subseteq J/I$. $\square$

The above result will allow us to control the quotient R/J^{n-1} in (7.3.2). In order to determine the behaviour of J^{n-1} as an R-module, we first need a definition.

Definition 7.3.11 Let I be an ideal of the ring R. We put $\mathrm{rad}(I) = \mathrm{rad}(R) \cap I$ and call it the *Jacobson radical of I*.

Clearly, this extends Definition 7.3.1 to non-unital rings. We collect some properties of this radical. Recall that we put $J = \mathrm{rad}(R)$. The first observation addresses the fact that a proper ideal has to be treated as an R-module, due to the lack of an identity.

Lemma 7.3.12 *For every ideal I of R, its radical is given by*

$$\mathrm{rad}(I) = \bigcap \{L' \mid L' \text{ maximal submodule of } {}_R I\}. \tag{7.3.3}$$

Proof Let us abbreviate the set on the right-hand side of (7.3.3) by J'. Note at first that, by definition,

$$\mathrm{rad}(I) = J \cap I = \bigcap \{L \cap I \mid L \text{ maximal left ideal of } R\}$$

$$= \bigcap \{L \cap I \mid L \text{ maximal left ideal of } R \text{ and } I \nsubseteq L\},$$

as $I \subseteq L$ is tantamount to $L \cap I = I$. Let $L \subseteq R$ be a maximal left ideal such that $I \nsubseteq L$. Then, $L + I = R$ and thus, by Theorem 3.1.3,

$$R/L = (L + I)/L \cong I/(L \cap I)$$

and since R/L is a simple left R-module, so is $I/(L \cap I)$. By Proposition 2.2.1, $L \cap I$ is a maximal submodule of I.

Now let $L' \subseteq I$ be a maximal submodule of $_R I$. Then $L' \subseteq L$ for some maximal left ideal L of R. Since $L' = L' \cap I \subseteq L \cap I$ the maximality of L' entails that, either $L' = L \cap I$, or $L \cap I = I$. Either case yields

$$\bigcap \{L \cap I \mid L \text{ maximal left ideal of } R \text{ and } I \nsubseteq L\} = J',$$

which was to show.

If $_R I$ does not contain any maximal submodules then $J' = I$. Moreover, every maximal left ideal L of R has to contain I so $I \subseteq J$ which implies that $J \cap I = I = J'$ in this case too. □

Combining this lemma with Proposition 6.1.10 we obtain the next result.

Corollary 7.3.13 *An ideal I of a left Artinian ring R is semisimple (in R-$\mathcal{M}od$) if and only if $\mathrm{rad}(I) = 0$.*

Proposition 7.3.14 *For every ideal I of a left Artinian ring R, we have $\mathrm{rad}(I) = JI$ and $I/\mathrm{rad}(I)$ is semisimple.*

Proof We prove the second statement first. From Theorem 3.1.3 we obtain

$$I/\mathrm{rad}(I) = I/(J \cap I) \cong (I + J)/J \hookrightarrow R/J \tag{7.3.4}$$

which together with Propositions 7.3.7 and 6.1.6 yields that $I/\mathrm{rad}(I)$ is semisimple.

Clearly, $JI \subseteq J \cap I$. By Exercise 7.4.10, $\mathrm{rad}(I/JI) = \mathrm{rad}(I)/JI$ so, in order to establish the reverse inclusion, it suffices to show that $\mathrm{rad}(I/JI)$ vanishes. Set $M = I/JI$; this is an Artinian left R-module, by Corollary 5.1.5. We can regard M as a left R/J-module (Exercise 2.7.3 and Corollary 7.3.3); since R/J is a semisimple ring (Corollary 7.3.6 and Proposition 7.3.7) it follows that M is a semisimple R/J-module (Theorem 6.1.12) and hence semisimple as an R-module and an R/JI-module (Exercise 2.7.3 again). By Corollary 7.3.13 above, $\mathrm{rad}(I/JI) = 0$ which was to prove. □

Remark 7.3.15 Let I be an ideal in a left Artinian ring R. The embedding (as rings) of $I/\mathrm{rad}(I)$ in (7.3.4) enables us to consider $I/\mathrm{rad}(I)$ as an ideal in R/J; call this $\bar{I}$. Then

$$\mathrm{rad}(I/\mathrm{rad}(I)) = \mathrm{rad}(\bar{I}) = \bar{I} \cap \mathrm{rad}(R/J) = 0,$$

by Corollary 7.3.6. Alternatively, this can be deduced from Exercise 7.4.10.

We are now in a position to provide a proof of the Hopkins–Levitzki theorem.

Proof of Theorem 5.3.2 Let R be a left Artinian ring and let $n \in \mathbb{N}$ be the least integer k such that $J^k = 0$ (Lemma 7.3.8). We prove that R is left Noetherian by induction on n. Let $n = 1$; then $J = 0$ and the assertion follows from Proposition 7.3.7 as every semisimple ring is left Noetherian (Corollary 6.2.5).

Now assume that $n > 1$ and, for every $1 \le k < n$ and every left Artinian ring S with $\mathrm{rad}(S)^k = 0$, we have S is left Noetherian. We will apply the short exact sequence (7.3.2) with $S = R/J^{n-1}$. By Corollary 5.1.5, S is left Artinian. By Lemma 7.3.10, $\mathrm{rad}(S) = \mathrm{rad}(R)/J^{n-1}$ and hence, $\mathrm{rad}(S)^{n-1} = \left(J/J^{n-1}\right)^{n-1} = 0$. Thus, by induction hypothesis, S is left Noetherian.

It remains to show that J^{n-1} is left Noetherian and then to apply Theorem 4.1.1 to complete the proof. Clearly, J^{n-1} is left Artinian and, by Proposition 7.3.14, $\mathrm{rad}(J^{n-1}) = J\,J^{n-1} = 0$ and J^{n-1} is semisimple (both as a left R/J- and a left R-module). By Theorem 6.2.4, J^{n-1} is left Noetherian, which completes the claim. $\qquad\square$

Remark 7.3.16 The above proof is inspired by the arguments in Sect. 15 of [2] where the theory of radicals of modules is discussed in detail.

Remark 7.3.17 The most commonly found proof of the Hopkins–Levitzki theorem uses composition series and relies on the Jordan–Hölder theorem. Let R be a ring. A *composition series of* a left R-module M is a descending chain of submodules

$$M = M_0 \supset M_1 \supset M_2 \supset \ldots \supset M_\ell = 0$$

such that each quotient M_{i-1}/M_i is simple. In this case, ℓ is the length of the composition series and the simple modules are called the composition factors. The Jordan–Hölder theorem states that any two composition series of a module have the same length and the composition factors are, up to permutation and isomorphism, uniquely determined. This allows one to introduce the length of a module as the length of a composition series, if such exists. It can be shown that a module has a finite composition series if and only if it is both Artinian and Noetherian. The composition series that one uses in a proof of the Hopkins–Levitzki theorem derive from the chain of inclusions

$$R \supset J \supset J^2 \supset \ldots \supset J^n = 0$$

as each quotient J^i/J^{i+1} is semisimple Artinian, so has a composition series itself. For details, we refer the reader to Sects. 2.5 and 3.3 in [7], in particular Theorem 3.3.5, and Chaps. 5 and 14 in [8].

The method using composition series yields versions of the Hopkins–Levitzki theorem for modules too; see, e.g., Theorem 4.15 in [24].

7.4 Exercises

Exercise 7.4.1 (Uniqueness Theorem for Semisimple Modules) Let $_R M \in$ R-$\mathcal{M}od$ for a unital ring R. Suppose that

$$_R M = \bigoplus_{i=1}^{n} {}_R M_i = \bigoplus_{j=1}^{m} {}_R M_j'$$

are two direct sum decompositions of $_R M$ with simple summands $_R M_i \leq {}_R M$ and $_R M_j' \leq {}_R M$. Then $n = m$ and there is a permutation π of $\{1, \ldots, n\}$ such that $_R M_i \cong {}_R M_{\pi(i)}'$ for each $1 \leq i \leq n$.

Exercise 7.4.2 Show that every non-zero Artinian module contains a simple sub-module.

Exercise 7.4.3 For a ring R, define its *socle* soc(R) as the sum of all minimal left ideals of R and if R has no minimal left ideals, put soc$(R) = 0$. Show that soc(R) is an ideal of R.

Exercise 7.4.4 For a ring R, an *essential* left ideal L of R is defined by the property that $L \cap J \neq 0$ for every non-zero left ideal J of R. Using Zorn's Lemma show that, for every left ideal I of R there is a left ideal I' of R such that $I \oplus I'$ is an essential left ideal.

Exercise 7.4.5 Let R be a ring with non-zero socle soc(R). Using Zorn's Lemma show that, for each $x \in R \setminus$ soc(R), there is an essential left ideal L of R containing soc(R) such that $x \notin L$. Use this to prove that soc(R) is equal to the intersection of all essential left ideals of R.

Exercise 7.4.6 Show that a unital ring R is semiprimitive if and only if it has a faithful semisimple left module.

Exercise 7.4.7 Let $R = R_1 \times R_2$ be the direct product of two rings R_1 and R_2. Show that $\mathrm{rad}(R) = \mathrm{rad}(R_1) \times \mathrm{rad}(R_2)$.

Exercise 7.4.8 A proper ideal P of a ring R is said to be *prime* if, whenever I_1, I_2 are ideals of R such that $I_1 I_2 \subseteq P$, we have $I_1 \subseteq P$ or $I_2 \subseteq P$, and it is called *left primitive* if it is the annihilator of a simple left R-module. Show that

  (i) every maximal ideal is left primitive;
 (ii) every left primitive ideal is prime;
(iii) if R is left Artinian then every prime ideal is maximal.

(Compare with Proposition 5.2.4 in Chap. 5.)

Exercise 7.4.9 A unital ring R is called *semiprime* if, for every ideal I of R, the condition $I^2 = 0$ entails $I = 0$. Prove the following analogue of Proposition 7.3.7: R is semisimple if and only if R is left Artinian and semiprime.

Exercise 7.4.10 Follow the argument in the proof of Lemma 7.3.10 to obtain the following extension of this lemma: Let I be an ideal in a unital ring R and let K be an ideal of R such that $K \subseteq \mathrm{rad}(I)$. Then $\mathrm{rad}(I/K) = \mathrm{rad}(I)/K$.

Exercise 7.4.11 Let $\rho \colon R \to S$ be a surjective ring homomorphism between the unital rings R and S. Show that $\rho(\mathrm{rad}(R)) \subseteq \mathrm{rad}(S)$. Give an example to show that equality may not always hold.

Exercise 7.4.12 Let R be a unital ring and let $n \in \mathbb{N}$. Using Proposition 7.1.1, show that $\mathrm{rad}(M_n(R)) = M_n(\mathrm{rad}(R))$.

Exercise 7.4.13 A unital ring R is called *von Neumann regular* if, for each $r \in R$, there is $s \in R$ such that $rsr = r$. Show that, if $_R M \in R\text{-}\mathcal{M}od$ is semisimple, then $\mathrm{End}_R(_R M)$ is von Neumann regular.

Exercise 7.4.14 Let R be a von Neumann regular ring. Show that every principal ideal of R (that is, an ideal which is generated by one element) is generated by an idempotent.

Exercise 7.4.15 Show that every von Neumann regular ring is semiprimitive. Let $C[0, 1]$ be the ring of all continuous real-valued functions on the unit interval $[0, 1]$. Prove that $C[0, 1]$ is semiprimitive but not von Neumann regular.

Exercise 7.4.16 Let R be a reduced ring in which every prime ideal is maximal. Show that R is von Neumann regular.

Exercise 7.4.17 An *algebra* over a field K is a ring A which at the same time is a K-vector space such that $\lambda(xy) = x(\lambda y)$ for all $\lambda \in K$ and $x, y \in A$. Suppose K is an algebraically closed field and that A is a finite-dimensional K-algebra. Let $_RM$ be a simple left A-module. Show that $\text{End}_A(_RM) \cong K$.

Exercise 7.4.18 Let $(I, \leq)$ be a directed set, and let $\{_RM_i \mid i \in I\}$ be a family of modules over a ring R. Suppose that, for $i \leq j$, there is a module homomorphism $f_{ji} \colon {}_RM_i \to {}_RM_j$ such that $f_{ii} = \text{id}_{|M_i}$ and, for $j \leq k$, $f_{ki} = f_{kj} \circ f_{ji}$. Then $\{(_RM_i, f_{ji}) \mid i, j \in I\}$ is called a *directed system of R-modules*. Let $_RM = \bigoplus_{i \in I} {}_RM_i$ be the direct sum of this family and let $_RN$ be the submodule of $_RM$ generated by $\{\iota_i(x_i) - \iota_j(f_{ji}(x_i)) \mid x_i \in {}_RM_i, \ i \leq j\}$, where $\iota_i \colon {}_RM_i \to {}_RM$, $i \in I$ is the canonical injection. Denote by $\varinjlim_I {}_RM_i$ the quotient module $_RM/_RN$. This is called the *direct limit* of the directed system of modules.

Define $\varepsilon_i \colon {}_RM_i \to \varinjlim_I {}_RM_i$ as the composition of ι_i followed on by the canonical quotient map $_RM \to {}_RM/_RN$. Show that $\{(\varinjlim_I {}_RM_i, \varepsilon_i) \mid i \in I\}$ has the following universal property:

It makes the right-hand triangle in the diagram below commutative and, whenever $\{(_RL, \eta_i) \mid i \in I\}$ consists of an R-module $_RL$ and module homomorphisms $\eta_i \colon {}_RM_i \to {}_RL$, $i \in I$ making the larger triangle commutative, then there is a unique R-module map $\varinjlim_I {}_RM_i \to {}_RL$ making the entire diagram commutative.

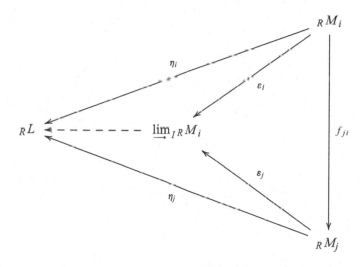

It should be obvious that the direct limit is uniquely determined up to isomorphism by this universal property.

Exercise 7.4.19 (Nakayama's Lemma)   Let I be a left ideal of the (unital) ring R. Show that the following three conditions are equivalent.

(i) $I \subseteq \mathrm{rad}(R)$;

(ii) for every $M \in R\text{-}mod$, $I \cdot M = M \Rightarrow M = 0$;

(iii) for all $_RN$, $_RM \in R\text{-}\mathcal{M}od$ with $N \subseteq M$ such that $_RM/_RN$ is finitely generated, $N + I \cdot M = M \Rightarrow N = M$.

Tensor Products of Modules

8

This chapter is devoted to a more sophisticated construction with modules, the so-called tensor product. It is extremely useful, though fairly abstract. Its main benefit lies in the fact that it allows us to convert bilinear mappings into homomorphisms of abelian groups. The relations between tensor products and homomorphism groups is fundamental and will lead us to the concept of adjoint functor in the later part of the chapter.

8.1   Tensor Product of Modules

Here, we will be dealing with left and right modules over a unital ring R at the same time.

Definition 8.1.1 Let $M_R \in \mathcal{M}od\text{-}R$ and $_RN \in R\text{-}\mathcal{M}od$. Let G be an abelian group. A mapping $\beta \colon M \times N \to G$ is said to be R-*balanced* if, for all $m, m_1, m_2 \in M, n, n_1, n_2 \in N$ and $r \in R$,

(i) $\beta(m_1 + m_2, n) = \beta(m_1, n) + \beta(m_2, n)$;
(ii) $\beta(m, n_1 + n_2) = \beta(m, n_1) + \beta(m, n_2)$;
(iii) $\beta(m \cdot r, n) = \beta(m, r \cdot n)$.

We denote the set of all such mappings by $B(M \times N, G)$.

Example 8.1.2 Let $R = \mathbb{R}$ and put $M = N = \mathbb{R}^3$ (as $\mathbb{R}$-vector spaces). We define

$$\beta \colon \mathbb{R}^3 \times \mathbb{R}^3 \to M_3(\mathbb{R}) \quad \text{via} \quad \begin{pmatrix} x_1 \\ x_2 \\ x_3 \end{pmatrix} \begin{pmatrix} y_1 & y_2 & y_3 \end{pmatrix} = \left(x_i y_j \right)_{1 \leq i, j \leq 3}.$$

© The Author(s), under exclusive license to Springer Nature Switzerland AG 2022
M. Mathieu, *Classically Semisimple Rings*,
https://doi.org/10.1007/978-3-031-14209-3_8

This 'vector product' is an example of an $\mathbb{R}$-balanced mapping into the abelian group of 3×3-matrices over $\mathbb{R}$.

Our approach to the tensor product is via its universal property.

Definition 8.1.3 A *tensor product* of the modules $M_R \in \mathscr{M}\!od\text{-}R$ and $_R N \in R\text{-}\mathscr{M}\!od$ over R is an abelian group $T(M, N)$ together with an R-balanced mapping $\tau : M \times N \to T(M, N)$ such that, for every abelian group G and each R-balanced mapping $\beta : M \times N \to G$, there is a unique $\mathbb{Z}$-module map $\hat{\beta} : T(M, N) \to G$ satisfying $\beta = \hat{\beta} \circ \tau$; that is, making the diagram below commutative

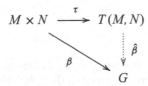

Theorem 8.1.4 *For all $M_R \in \mathscr{M}\!od\text{-}R$ and $_R N \in R\text{-}\mathscr{M}\!od$ over a unital ring R there is a tensor product which is unique up to isomorphism.*

Notation As a consequence of Theorem 8.1.4, we shall denote this unique tensor product by $M \otimes_R N$ and the associated mapping τ by $(x, y) \mapsto x \otimes y$.

Proof We will prove the uniqueness first. Suppose that both (T, τ) and (S, σ) satisfy the universal property of (8.1.3). By applying it to σ and to τ, respectively we obtain the following two commutative diagrams

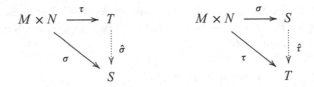

We claim that the group homomorphisms $\hat{\sigma}$ and $\hat{\tau}$ are inverses of each other so that T and S are isomorphic. We have

$$\hat{\sigma}\,\hat{\tau}\,\sigma = \hat{\sigma}\,\tau = \sigma \quad \text{and} \quad \hat{\tau}\,\hat{\sigma}\,\tau = \hat{\tau}\,\sigma = \tau.$$

As a result, we obtain the following commutative diagrams

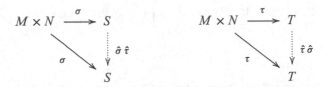

The uniqueness hypothesis in (8.1.3) implies that $\hat{\sigma}\,\hat{\tau} = \mathrm{id}_S$ and $\hat{\tau}\,\hat{\sigma} = \mathrm{id}_T$ which was to prove.

We next show the existence of a tensor product. Let F be the free $\mathbb{Z}$-module with basis $M \times N$ and consider $M \times N$ as a submodule of F. Recall that this means, each element of F is a unique $\mathbb{Z}$-linear combination $\sum_{i=1}^{k} n_i (x_i, y_i)$ with $n_i \in \mathbb{Z}$, $x_i \in M$, $y_i \in N$ and $k \in \mathbb{N}$; compare Sect. 2.5. Let K be the $\mathbb{Z}$-submodule of F generated by elements of the form

$$(x_1 + x_2, y) - (x_1, y) - (x_2, y),$$

$$(x, y_1 + y_2) - (x, y_1) - (x, y_2),$$

$$(x \cdot r, y) - (x, r \cdot y),$$

where $x, x_1, x_2 \in M$, $y, y_1, y_2 \in N$ and $r \in R$. Put $T = F/K$ and $\pi = \pi_K$, the canonical epimorphism. We set

$$\tau : M \times N \to T, \quad \tau(x, y) = \pi(x, y)$$

and verify the conditions in (8.1.3).

Let $x, x_1, x_2 \in M$, $y \in N$, $r \in R$; then

$$\tau(x_1 + x_2, y) = \pi(x_1 + x_2, y) = (x_1 + x_2, y) + K$$
$$= (x_1, y) + K + (x_2, y) + K = \tau(x_1, y) + \tau(x_2, y)$$

by the definition of K. Similarly, τ is additive in the second variable. Moreover,

$$\tau(x \cdot r, y) = \pi(x \cdot r, y) = (x \cdot r, y) + K = (x, r \cdot y) + K = \tau(x, r \cdot y).$$

This shows that τ is R-balanced.

Now let $\beta : M \times N \to G$ be an R-balanced mapping into an abelian group G. As F is free, there is a unique extension of β to a $\mathbb{Z}$-module map $\tilde{\beta} : F \to G$ and we put $\hat{\beta} \circ \pi = \tilde{\beta}$. This is a $\mathbb{Z}$-module map provided it is well defined. To this end, take $x, x_1, x_2 \in M$, $y \in N$, $r \in R$ and observe that

$$\tilde{\beta}\big((x_1 + x_2, y) - (x_1, y) - (x_2, y)\big) = \beta(x_1 + x_2, y) - \beta(x_1, y) - \beta(x_2, y)$$
$$= \beta(x_1 + x_2 - x_1 - x_2, y) = 0$$

and similarly for the second variable. Moreover,

$$\tilde{\beta}\big((x \cdot r, y) - (x, r \cdot y)\big) = \beta(x \cdot r, y) - \beta(x, r \cdot y) = 0.$$

This shows that $K = \ker \pi \subseteq \ker \tilde{\beta}$ and so $\pi(f) = 0$ implies $\tilde{\beta}(f) = 0$ for each $f \in F$. As a result, $\hat{\beta}$ is well defined.

Finally, $\hat{\beta} \circ \tau = \hat{\beta} \circ \pi_{|M \times N} = \beta$ yields the universal property. □

Remarks 8.1.5

1. For any $M_R \in \mathcal{M}od\text{-}R$ and $_RN \in R\text{-}\mathcal{M}od$, every element in $M \otimes_R N$ as constructed in (8.1.4) can be written as

$$\sum_{i=1}^{k} n_i(x_i, y_i) + K,$$

   but $n_i(x_i, y_i) + K = (n_i x_i, y_i) + K$ and $n_i x_i \in M_R$ so we can simply write $\sum_{i=1}^{k}(x_i, y_i) + K$ which we will denote by $\sum_{i=1}^{k} x_i \otimes y_i$.
2. If the ring R is commutative, every left R-module can be considered as a right R-module in a canonical way and vice versa; see Exercise 8.4.1. In this case, $M \otimes_R N$ carries a natural module structure. It follows that, if $\{x_i \mid i \in I\}$ and $\{y_j \mid j \in J\}$ are sets of generators of M_R and $_RN$, respectively then $\{x_i \otimes y_j \mid (i, j) \in I \times J\}$ is a set of generators of $M \otimes_R N$. In particular, the tensor product of two finitely generated modules over a commutative ring is finitely generated.

The tensor product is rather well behaved and compatible with other constructions with modules. As an example, we mention the result below.

Proposition 8.1.6 *Let* $M_R, M_R{}' \in \mathcal{M}od\text{-}R$ *and* $_RN, {}_RL \in R\text{-}\mathcal{M}od$. *Then*

   (i) $(M \oplus M') \otimes_R N \cong M \otimes_R N \oplus M' \otimes_R N$;
  (ii) $M \otimes_R N \cong N \otimes_R M$ *provided* R *is commutative*;
 (iii) $(M \otimes_R N) \otimes_R L \cong M \otimes_R (N \otimes_R L)$ *provided* R *is commutative*.

Proof We leave the verification of (i) and (iii) to the reader; see the Exercises. Note that, if R is commutative, the tensor product carries a natural R-module structure and therefore, it makes sense to iterate the construction as in (iii).

For illustration, we prove assertion (ii). Suppose R is commutative so that we can consider the modules M and N as modules on the left and on the right. We have the following commutative diagram

$$
\begin{array}{ccc}
M \times N & \xrightarrow{\ \iota\ } & N \times M \\
\Big\downarrow{\scriptstyle \tau_1} & & \Big\downarrow{\scriptstyle \tau_2} \\
M \otimes_R N & \underset{\hat{\tau}_1}{\overset{\hat{\tau}_2}{\rightleftarrows}} & N \otimes_R M
\end{array}
$$

in which the 'flip' $\iota\colon (x, y) \mapsto (y, x)$ is clearly a bilinear isomorphism.

The additive maps $\hat{\tau}_1$, $\hat{\tau}_2$ are given by the universal property of the tensor product and satisfy $\hat{\tau}_2 \circ \tau_1 = \tau_2 \circ \iota$ and $\hat{\tau}_1 \circ \tau_2 = \tau_1 \circ \iota^{-1}$. Therefore

$$\tau_2 \circ \iota = \hat{\tau}_2 \circ \tau_1 = \hat{\tau}_2 \circ \hat{\tau}_1 \circ \tau_2 \circ \iota$$

and

$$\tau_1 \circ \iota^{-1} = \hat{\tau}_1 \circ \tau_2 = \hat{\tau}_1 \circ \hat{\tau}_2 \circ \tau_1 \circ \iota^{-1}$$

which gives $\hat{\tau}_2 \circ \hat{\tau}_1 = \mathrm{id}_{|N \otimes_R M}$ and $\hat{\tau}_1 \circ \hat{\tau}_2 = \mathrm{id}_{|M \otimes_R N}$, respectively, where, again, we applied the universal property. Consequently, $M \otimes_R N \cong N \otimes_R M$ as abelian groups.

Since, for all $x \in M$, $y \in N$,

$$\hat{\tau}_2(r \cdot x \otimes y) = \hat{\tau}_2 \circ \tau_1(r \cdot x, y) = \tau_2 \circ \iota(r \cdot x, y) = \tau_2(y, r \cdot x)$$
$$= y \otimes r \cdot x = r \cdot (y \otimes x)$$
$$= r \cdot \tau_2(y, x) = r \cdot \tau_2 \circ \iota(x, y) = r \cdot \hat{\tau}_2 \circ \tau_1(x, y)$$
$$= r \cdot \hat{\tau}_2(x \otimes y)$$

and similarly for τ_1, we see that this isomorphism is indeed an isomorphism of R-modules. □

Unless M is an R-bimodule, we cannot endow $M \otimes_R N$ with a left module structure. There is the following important special case, however.

Proposition 8.1.7 *Let R be a unital ring, and let $_R N \in R\text{-}\mathcal{M}od$. Then $R \otimes_R N \cong {}_R N$ as left R-modules.*

Proof We consider R as standard right R-module R_R. Define $\beta: R \times N \to N$ by $\beta(a, x) = a \cdot x$, $a \in R$, $x \in N$. Then β is R-balanced:

$$\beta(a_1 + a_2, x) = (a_1 + a_2) \cdot x = a_1 \cdot x + a_2 \cdot x = \beta(a_1, x) + \beta(a_2, x);$$
$$\beta(a, x + y) = a \cdot (x + y) = a \cdot x + a \cdot y = \beta(a, x) + \beta(a, y);$$
$$\beta(ar, x) = (ar) \cdot x = a \cdot (r \cdot x) = \beta(a, r \cdot x)$$

for all $a, a_1, a_2, r \in R$ and $x, y \in N$.

Let $\hat{\beta}: R \otimes_R N \to N$ be the unique group homomorphism satisfying $\hat{\beta}(a \otimes x) = a \cdot x$ given by (8.1.4). In this case, $R \otimes_R N$ is also a left R-module in a canonical way and $\hat{\beta}$ is an R-module map:

$$r \cdot \hat{\beta}(a \otimes x) = r \cdot (a \cdot x) = (ra) \cdot x = \hat{\beta}(ra \otimes x)$$

which extends to sums via additivity.

Let $x \in N$; then $x = 1 \cdot x = \hat{\beta}(1 \otimes x)$, thus $\hat{\beta}$ is surjective. Put

$$\gamma : N \to R \otimes_R N, \quad x \mapsto 1 \otimes x.$$

Then, for each $\sum_i a_i \otimes x_i \in R \otimes_R N$,

$$\gamma \hat{\beta}\Big(\sum_i a_i \otimes x_i\Big) = \gamma\Big(\sum_i a_i \cdot x_i\Big) = 1 \otimes \sum_i a_i \cdot x_i = \sum_i a_i \otimes x_i.$$

Therefore $\gamma \hat{\beta} = \mathrm{id}_{R \otimes_R N}$ and $\hat{\beta}$ is injective. The statement follows.   □

Our next aim is to investigate the relation between the tensor product and homomorphism groups. To this end, we first record the following special situation.

Proposition 8.1.8 *Let R be a unital ring, and let $_R N \in R\text{-}\mathscr{M}od$. Then* $\mathrm{Hom}_R(R, {}_R N) \cong {}_R N$ *as left R-modules.*

Proof Let $f \in \mathrm{Hom}_R(R, {}_R N)$; then, for $r \in R$, $r \cdot f$ defined by $(r \cdot f)(a) = f(ar)$, $a \in R$ is in $\mathrm{Hom}_R(R, {}_R N)$ so $\mathrm{Hom}_R(R, {}_R N)$ is a left R-module. Define $\varphi : \mathrm{Hom}_R(R, {}_R N) \to {}_R N$ by $\varphi(f) = f(1)$, $f \in \mathrm{Hom}_R(R, {}_R N)$; we have

$$\varphi(f_1 + f_2) = (f_1 + f_2)(1) = f_1(1) + f_2(1) = \varphi(f_1) + \varphi(f_2)$$

and

$$\varphi(r \cdot f) = (r \cdot f)(1) = f(r) = r \cdot f(1) = r \cdot \varphi(f)$$

for all $r \in R$ and $f, f_1, f_2 \in \mathrm{Hom}_R(R, {}_R N)$. Thus φ is an R-module map. Take $f \in \ker \varphi$; then $f(r) = r \cdot f(1) = 0$ for every $r \in R$ so $f = 0$. Hence φ is injective. Take $y \in {}_R N$. Define $f \in \mathrm{Hom}_R(R, {}_R N)$ by $f(a) = a \cdot y$. As $y = f(1) = \varphi(f)$, φ is surjective and altogether an isomorphism.   □

Combining the last two propositions we have, as abelian groups,

$$\mathrm{Hom}_R(R \otimes_R M, {}_R N) \cong \mathrm{Hom}_R({}_R M, {}_R N) \cong \mathrm{Hom}_R({}_R M, \mathrm{Hom}_R(R, {}_R N))$$

for all $_R M, {}_R N \in R\text{-}\mathscr{M}od$. We shall now see that a similar isomorphism can be established in general once we allow ourselves to work with bimodules.

Theorem 8.1.9 *Let S and R be unital rings, let $_R M \in R\text{-}\mathscr{M}od$, $_S N \in S\text{-}\mathscr{M}od$ and $_S U_R \in S\text{-}\mathscr{M}od\text{-}R$. Then*

$$\mathrm{Hom}_S({}_S U \otimes_R M, {}_S N) \cong \mathrm{Hom}_R({}_R M, \mathrm{Hom}_S({}_S U, {}_S N)).$$

Proof In order to define a group isomorphism

$$\varphi \colon \mathrm{Hom}_S \left({_S U} \otimes_R M, {_S N} \right) \to \mathrm{Hom}_R \left({_R M}, \mathrm{Hom}_S \left({_S U}, {_S N} \right) \right),$$

take $f \in \mathrm{Hom}_S \left({_S U} \otimes_R M, {_S N} \right)$. Fix $x \in M$ and put $f_x \colon U \to N$, $f_x(u) = f(u \otimes x)$ for each $u \in U$. In this way, we get an S-module map f_x:

$$f_x(u_1 + u_2) = f((u_1 + u_2) \otimes x)$$
$$= f(u_1 \otimes x + u_2 \otimes x) = f(u_1 \otimes x) + f(u_2 \otimes x)$$
$$= f_x(u_1) + f_x(u_2),$$
$$f_x(s \cdot u) = f(s \cdot u \otimes x) = s \cdot f(u \otimes x)$$
$$= s \cdot f_x(u)$$

for all $u, u_1, u_2 \in U$ and $x \in M$. Hence $f_x \in \mathrm{Hom}_S \left({_S U}, {_S N} \right)$ and we define $\varphi(f)(x) = f_x$, $x \in M$. For $x, y \in M$, we have

$$\varphi(f)(x + y) = f_{x+y} = f_x + f_y = \varphi(f)(x) + \varphi(f)(y)$$

since $f_{x+y}(u) = f_x(u) + f_y(u)$ for all $u \in U$. Moreover,

$$f_{r \cdot x}(u) = f(u \otimes r \cdot x) = f(u \cdot r \otimes x) = f_x(u \cdot r) = (r \cdot f_x)(u) \quad (u \in U)$$

where the last equality comes from the definition of $r \cdot f_x$. Consequently, for $x \in M$ and $r \in R$, we have

$$\varphi(f)(r \cdot x) = f_{r \cdot x} = r \cdot f_x = r \cdot \varphi(f)$$

and thus $\varphi(f)$ is an R-module map.

Next we show that φ is a group homomorphism. Take $x \in M$, $f, g \in \mathrm{Hom}_S \left({_S U} \otimes_R M, {_S N} \right)$. For all $u \in U$,

$$(f + g)_x(u) = (f + g)(u \otimes x) = f(u \otimes x) + g(u \otimes x) = f_x(u) + g_x(u)$$

and thus

$$\varphi(f + g)(x) = (f + g)_x = f_x + g_x = \varphi(f)(x) + \varphi(g)(x) = (\varphi(f) + \varphi(g))(x).$$

Take $f \in \ker \varphi$. Then $f_x(u) = f(u \otimes x) = 0$ for all $x \in M$ and therefore $f = 0$.

We now define a right inverse of φ to show that φ is surjective, which will complete the proof. Let $h \in \mathrm{Hom}_R \left({_R M}, \mathrm{Hom}_S \left({_S U}, {_S N} \right) \right)$. Put

$$\beta_h \colon U \times M \to N, \quad \beta_h(u, x) = h(x)(u)$$

whenever $u \in U$ and $x \in M$. It is evident that β_h is additive in the first and in the second variable separately. In addition, for $u \in U, x \in M$ and $r \in R$,

$$\beta_h(u \cdot r, x) = h(x)(u \cdot r) = (r \cdot h(x))(u) = h(r \cdot x)(u) = \beta_h(u, r \cdot x)$$

so that β_h is R-balanced. The universal property of the tensor product yields a unique group homomorphism $\psi(h)\colon {}_S U \otimes_R M \to N$ with the property $\psi(h)(u \otimes x) = \beta_h(u, x)$ for all $u \in U, x \in M$. From

$$\big(\varphi(\psi(h))(x)\big)(u) = \psi(h)_x(u) = \psi(h)(u \otimes x) = h(x)(u) \qquad (u \in U)$$

we obtain $\psi(h)_x = h(x)$ for all $x \in M$ and therefore $\varphi(\psi(h)) = h$. As a result, $\varphi \circ \psi = \mathrm{id}$ which was to prove.   □

In categorical language the above theorem states that the functors $U \otimes_R -$ and $\mathrm{Hom}_S(U, -)$ are adjoint functors; compare Sect. 8.3 below.

We next introduce the tensor product of module maps.

Proposition 8.1.10 *Let R be a unital ring. Let $M_R, M_R' \in \mathcal{M}od\text{-}R$ and ${}_R N, {}_R N' \in R\text{-}\mathcal{M}od$. For every pair of module maps $f \in \mathrm{Hom}_R(M_R, M_R')$, $g \in \mathrm{Hom}_R({}_R N, {}_R N')$ there is a unique $\mathbb{Z}$-module map*

$$f \otimes g \colon M \otimes_R N \longrightarrow M' \otimes_R N'$$

such that $(f \otimes g)(x \otimes y) = f(x) \otimes g(y)$ for all $x \in M, y \in N$.

Proof Define $\beta \colon M \times N \to M' \otimes_R N'$ by $\beta(x, y) = f(x) \otimes g(y), x \in M, y \in N$. Then

$$\beta(x, y_1 + y_2) = f(x) \otimes g(y_1 + y_2) = f(x) \otimes (g(y_1) + g(y_2))$$
$$= f(x) \otimes g(y_1) + f(x) \otimes g(y_2) = \beta(x, y_1) + \beta(x, y_2)$$

for all $x \in M, y_1, y_2 \in N$. Similarly, we show that β is additive in the first variable. Furthermore, for $x \in M, y \in N$ and $r \in R$,

$$\beta(x \cdot r, y) = f(x \cdot r) \otimes g(y) = f(x) \cdot r \otimes g(y) = f(x) \otimes r \cdot g(y)$$
$$= f(x) \otimes g(r \cdot y) = \beta(x, r \cdot y).$$

Hence β is R-balanced and, by the universal property of the tensor product, we obtain a unique extension to a $\mathbb{Z}$-module map $M \otimes_R N \to M' \otimes_R N'$; we denote this map by $f \otimes g$.   □

8.2   Tensor Product of Algebras

In this section, K will denote a field and A, B, C will be unital K-algebras. That is, A is a K-vector space and carries an associative multiplication $(x, y) \mapsto xy$ such that $\lambda(xy) = x(\lambda y)$ for all $x, y \in A$ and the two distributivity laws hold. Furthermore, there is a multiplicative identity which we will denote by 1.

Our aim is to endow the module tensor product $A \otimes_K B$ with an algebra structure and study the properties of this algebra. To this end, we first amend the definition of a K-balanced mapping in a natural way. A mapping $\beta \colon A \times B \to C$ will be called K-bilinear if, for all $a, a_1, a_2 \in A$, $b, b_1, b_2 \in B$ and $\lambda \in K$,

 (i) $\beta(a_1 + a_2, b) = \beta(a_1, b) + \beta(a_2, b)$;
 (ii) $\beta(a, b_1 + b_2) = \beta(a, b_1) + \beta(a, b_2)$;
 (iii) $\beta(\lambda a, b) = \beta(a, \lambda b) = \lambda \beta(a, b)$.

Note that the only difference to the original definition of an R-balanced mapping (p. 93) is in the last identity of the last line above which is possible as C is a vector space too.

A slight modification of the construction of the module tensor product in Theorem 8.1.4 yields a K-vector space $A \otimes_K B$ which has the universal property with respect to K-bilinear mappings as defined above, that is, results in a *linear* mapping $\hat{\beta}$ from $A \otimes_K B$ into a K-vector space E.

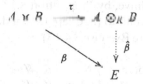

There is a canonical way of turning this vector space tensor product into a K-algebra.

Definition 8.2.1 Let A and B be unital algebras over the field K. The tensor product algebra $A \otimes_K B$ is the vector space tensor product endowed with the unique associative product satisfying $(a \otimes b)(c \otimes d) = ac \otimes bd$ for all $a, c \in A$ and $b, d \in B$.

It should by now be evident how to define the product alluded to above: we first have to linearise the K-bilinear mapping

$$(A \times B) \times (A \times B) \longrightarrow (A \times B) \otimes_K (A \times B), \quad \big((a, b), (c, d)\big) \longmapsto (a, b) \otimes (c, d),$$

where $A \times B$ carries the canonical algebra structure. In a second step we linearise the bilinear mapping

$$(A \times B) \otimes_K (A \times B) \longrightarrow A \otimes_K B, \quad (a, b) \otimes (c, d) \longmapsto ac \otimes bd$$

in order to obtain the desired multiplication.

Remark 8.2.2 Just as in Remark 8.1.5 we observe that, whenever $\{a_i \mid i \in I\}$ is a basis of A and $\{b_j \mid j \in J\}$ is a basis of B, then $\{a_i \otimes b_j \mid (i, j) \in I \times J\}$ is a basis of the vector space $A \otimes_K B$. Therefore, if $\dim_K A = n < \infty$ and $\dim_K B = m < \infty$, the dimension of $A \otimes_K B$ is finite, in fact, it is equal to nm.

The mappings $a \mapsto a \otimes 1$ and $b \mapsto 1 \otimes b$ are injective algebra homomorphisms from A into $A \otimes_K B$ and from B into $A \otimes_K B$, respectively, and their images, denoted by $A \otimes 1$ and $1 \otimes B$, respectively, commute with each other and generate $A \otimes_K B$.

8.3 Adjoint Functors

Let R be a unital ring. For fixed $M_R \in \mathcal{M}od\text{-}R$, $M \otimes_R -$ is a covariant functor from $R\text{-}\mathcal{M}od$ into $\mathcal{AG}r$, the category of abelian groups (use Exercises 8.4.11 and 8.4.12). Likewise, for fixed $_R N \in R\text{-}\mathcal{M}od$, $- \otimes_R N$ is a covariant functor from $\mathcal{M}od\text{-}R$ into $\mathcal{AG}r$. These functors are additive, by Proposition 8.1.6, and right exact. (To show this requires some work; we refer to [8, Chap. 15] for details.) However they are in general not left exact (compare Exercise 8.4.13).

Definition 8.3.1 A left R-module $_R N \in R\text{-}\mathcal{M}od$ is called *flat* if for every monomorphism $f \in \mathrm{Hom}_R(M_R{}', M_R)$ in $\mathcal{M}od\text{-}R$ the induced $\mathbb{Z}$-module map $f \otimes \mathrm{id} \colon M' \otimes_R N \to M \otimes_R N$ is a monomorphism in $\mathcal{AG}r$.

Proposition 8.3.2 *For every unital ring R, the left module $_R R$ is flat.*

Proof A right-handed version of Proposition 8.1.7 states that $M_R \cong M_R \otimes R$ for every right R-module M_R, and the analogous proof shows that $\gamma_M \colon x \mapsto x \otimes 1$ is an isomorphism. Given a monomorphism $f \in \mathrm{Hom}_R(M_R{}', M_R)$ in $\mathcal{M}od\text{-}R$ we thus have $f \otimes \mathrm{id} = \gamma_M \circ f \circ \gamma_{M'}^{-1}$ is a monomorphism. $\qquad\square$

Generalising Proposition 8.1.6 one can show that $\left(\oplus_{i \in I} M_i\right) \otimes_R N \cong \oplus_{i \in I} M_i \otimes_R N$ for every family $\{M_i \mid i \in I\}$ in $\mathcal{M}od\text{-}R$ and each module $_R N \in R\text{-}\mathcal{M}od$. This is done of course by showing that $\left(\oplus_{i \in I} M_i\right) \otimes_R N$ is a coproduct in $\mathcal{AG}r$ of the family $\{M_i \otimes_R N \mid i \in I\}$. Together with Proposition 8.3.2 this implies immediately that every free left R-module is flat. Using this fact together with Proposition 8.1.6 and Exercise 6.4.7 gives us the following corollary.

Corollary 8.3.3 *Every projective left module over a unital ring is flat.*

Recall the concept of a von Neumann regular ring R: for each $r \in R$ there is $s \in R$ such that $rsr = r$ (Exercise 7.4.13). It turns out that a unital ring R is von Neumann regular if and only if every left R-module is flat; see, e.g., [8, Theorem 15.22]. This requires a more detailed study of the finitely generated left ideals in a von Neumann regular ring.

By definition, a module $_R N \in R\text{-}\mathcal{M}od$ is flat if and only if the functor $- \otimes_R N$ is exact (compare Sect. 4.2.3). We shall now discuss the "Fundamental theorem of tensor products" (Theorem 8.1.9) from the point of view of adjoint functors.

Definition 8.3.4 Let $\mathscr{C}$ and $\mathscr{D}$ be categories and suppose $F: \mathscr{C} \to \mathscr{D}$ and $G: \mathscr{D} \to \mathscr{C}$ are covariant functors. Suppose there exists a natural equivalence

$$\eta = \eta_{AB}: \text{Mor}_{\mathscr{D}}(F(A), B) \longrightarrow \text{Mor}_{\mathscr{C}}(A, G(B)), \quad A \in \text{ob}(\mathscr{C}), \ B \in \text{ob}(\mathscr{D})$$

of functors $\mathscr{C}^{\text{op}} \times \mathscr{D} \to \mathcal{S}$. Then F is called *left adjoint* to G and G is called *right adjoint* to F. Moreover, η is called the *adjugant*.

The naturality of η in the above definition can be made explicit as follows. For all $f \in \text{Mor}_{\mathscr{C}}(A', A)$, $g \in \text{Mor}_{\mathscr{D}}(B, B')$ and $h \in \text{Mor}_{\mathscr{D}}(F(A), B)$, we have

$$\eta(g \circ h \circ F(f)) = G(g) \circ \eta(h) \circ f.$$

Let us look at some examples of adjoint functors we already encountered.

Examples 8.3.5

1. Let $F: \mathcal{S} \to R\text{-}\mathcal{M}od$ be the 'free functor' (compare Sect. 2.5) and let $G: R\text{-}\mathcal{M}od \to \mathcal{S}$ be the forgetful functor. Then F is left adjoint to G and G is right adjoint to F by Theorem 2.5.3. This example extends to any concrete category which has free objects in the sense of Sect. 2.6.1.
2. Theorem 8.1.9 tells us (in essence) that, whenever R, S are unital rings and $_S U_R \in S\text{-}\mathcal{M}od\text{-}R$, then the functor $_S U \otimes_R -$ is left adjoint to the functor $\text{Hom}_S (_S U, -)$.
3. In a similar vein, if R and S are unital rings and $\rho: R \to S$ is a unital ring homomorphism then, considering S as a right R-module (see Exercise 1.3.3), the 'extension-of-scalars' functor $S \otimes_R -: R\text{-}\mathcal{M}od \to S\text{-}\mathcal{M}od$ is left adjoint to the 'restriction-of-scalars' functor $S\text{-}\mathcal{M}od \to R\text{-}\mathcal{M}od$.
4. Let R be a unital ring and let G be a (multiplicative) group. The construction of the group ring (see Examples 1.2 (ix)) provides us with a functor $R[-]$ from the category of groups into the category of unital rings. It is left adjoint to the functor in the opposite direction which associates to a unital ring R its group of units R^*.
5. We return to Example 3.2.4 from our present point of view. For a commutative unital C^*-algebra A, its Gelfand space $\Delta(A)$ is homeomorphic to the space of all multiplicative linear functionals on A equipped with the weak $*$-

topology; thus the functor Δ can also be viewed as the representable functor $\mathrm{Mor}_{\mathscr{AC}_1^*}(-, \mathbb{C}) \colon \mathscr{AC}_1^* \longrightarrow \mathscr{Comp}$. In the other direction, the functor C associates with a compact Hausdorff space X the morphism set $\mathrm{Mor}_{\mathscr{Comp}}(X, \mathbb{C})$. Therefore we obtain the adjunction

$$\mathrm{Mor}_{\mathscr{Comp}}(\mathrm{Mor}_{\mathscr{AC}_1^*}(A, \mathbb{C}), Y) \cong \mathrm{Mor}_{\mathscr{AC}_1^*}(\mathrm{Mor}_{\mathscr{Comp}}(Y, \mathbb{C}), A),$$

where we have to take care of the fact the both functors are contravariant.

Adjoint functors are ubiquitous in Mathematics; for a detailed study see, for instance, Sect. II.7 in [21]. They have pleasant properties; e.g., left adjoint functors preserve colimits while right adjoint functors preserve limits. As an illustration of the arguments, we prove a special case of the latter statement. A pullback in a general category, if it exists, is defined by the universal property as stated in Exercise 4.3.13.

Proposition 8.3.6 *Suppose that* $\mathsf{G} \colon \mathscr{D} \to \mathscr{C}$ *is a covariant functor which has a left adjoint (so it is itself a right adjoint functor). Then* G *preserves products and pullbacks.*

Proof Let $\{B_i \mid i \in I\}$ be a family of objects in $\mathscr{D}$ and let $(B, p_i)_{i \in I}$ denote their product (assuming it exists). To show that $(\mathsf{G}(B), \mathsf{G}(p_i))_{i \in I}$ is the product of the family $\{\mathsf{G}(B_i) \mid i \in I\}$ in $\mathscr{C}$, let $f_i \colon A \to \mathsf{G}(B_i)$, $i \in I$ be morphisms in $\mathscr{C}$. Let $\eta \colon \mathsf{F} \to \mathsf{G}$ be an adjugant and let ξ be its inverse. For each i, $\xi(f_i) \colon \mathsf{F}(A) \to B_i$ so that there exists a unique morphism $g \colon \mathsf{F}(A) \to B$ in $\mathscr{D}$ such that $p_i \circ g = f_i$ for all $i \in I$. It follows that

$$\mathsf{G}(p_i) \circ \eta(g) = \eta(p_i \circ g) = f_i \quad (i \in I),$$

and $\eta(g)$ is the unique morphism with this property as every morphism $f' \colon A' \to \mathsf{G}(B)$ is of the form $f' = \eta(g')$ for some g' in $\mathrm{mor}(\mathscr{D})$.

To prove the second assertion,

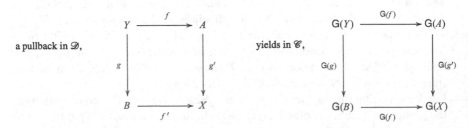

a pullback in $\mathscr{D}$,   yields in $\mathscr{C}$,

which we need to show is a pullback. Suppose $h \colon Z \to \mathsf{G}(A)$ and $k \colon Z \to \mathsf{G}(B)$ satisfy $\mathsf{G}(g') \circ h = \mathsf{G}(f') \circ k$. Upon applying ξ we obtain $g' \circ \xi(h) = f' \circ \xi(k)$ and thus there exists a unique morphism $\rho \colon \mathsf{F}(Z) \to Y$ such that $f \circ \rho = \xi(h)$ and

$g \circ \rho = \xi(k)$. Now applying η yields $G(f) \circ \eta(\rho) = h$ and $G(g) \circ \eta(\rho) = k$ and, as for products, $\eta(\rho)$ is the unique morphism satisfying these identities. □

8.4 Exercises

Exercise 8.4.1 Let R be a commutative ring and let $_R M \in R\text{-}\mathcal{M}od$. Show that M becomes a right R-module in a natural way by defining $x \cdot r = r \cdot x$, $x \in M, r \in R$. As a result, when considering modules over commutative rings, one usually does not specify "left" or "right".

Exercise 8.4.2 Prove the two outstanding statements in Proposition 8.1.6.

Exercise 8.4.3 Show that, for every finite abelian group G, $\mathbb{Q} \otimes_{\mathbb{Z}} G = 0$.

Exercise 8.4.4 Let K be a field and let G, H be groups. Prove that $K[G] \otimes_K K[H] \cong K[G \times H]$ as K-algebras.

Exercise 8.4.5 Let K be a field and let G be a group. Take $M, N \in \mathcal{M}od\text{-}K[G]$ and consider M and N as K-vector spaces in order to form their tensor product $M \otimes_K N$. Show that $(m \otimes n)g \mapsto (mg) \otimes (ng)$ defines a right $K[G]$-module structure on $M \otimes_K N$.

Exercise 8.4.6 Let K be a field and let A be a unital K-algebra. Show that, for all $n \in \mathbb{N}$, $M_n(A) \cong M_n(K) \otimes_K A$.

Exercise 8.4.7 Let V and W be vector spaces over the field K. Their *vector space tensor product* $V \otimes W$ is defined as the module tensor product $V \otimes_K W$. Show that, for every $z \in V \otimes W$, there exist $n \in \mathbb{N}$, linearly independent vectors $v_1, \ldots, v_n \in V$ and linearly independent vectors $w_1, \ldots, w_n \in W$ such that $z = \sum_{i=1}^{n} v_i \otimes w_i$.

Exercise 8.4.8 For a K-vector space V, let $L(V)$ denote the K-algebra of all linear mappings from V to itself. Suppose V and W are K-vector spaces. Show that, for each $S \in L(V)$, $T \in L(W)$, the function $\gamma_{S,T}$ defined by $V \times W \to V \otimes W$, $\gamma_{S,T}(v, w) = Sv \otimes Tw$ for all $v \in V$, $w \in W$ yields a linear mapping $\Gamma_{S,T} : V \otimes W \to V \otimes W$ such that $\Gamma_{S,T}(v \otimes w) = Sv \otimes Tw$ and from this, a bilinear mapping

$$\gamma : L(V) \times L(W) \longrightarrow L(V \otimes W), \quad \gamma(S, T) = \Gamma_{S,T} \quad (S \in L(V), T \in L(W))$$

is obtained. Use the universal property of the algebra tensor product to obtain a unique homomorphism

$$\varphi : L(V) \otimes L(W) \longrightarrow L(V \otimes W)$$

such that $\varphi(S \otimes T) = \Gamma_{S,T}$ for all $S \in L(V)$, $T \in L(W)$. Use the previous exercise to show that φ is injective, and refer to Remark 8.1.5 to prove that φ is surjective provided both V and W are finite dimensional.

Exercise 8.4.9 Let A and B be finite-dimensional semisimple algebras over an algebraically closed field K. Use the isomorphism from the previous exercise together with the Artin–Wedderburn theorem to prove that $A \otimes_K B$ is semisimple.

Exercise 8.4.10 With the assumptions and notation as in Proposition 8.1.10 show that $f \otimes g$ is surjective if both f and g are surjective. Does an analogous statement hold for injectivity?

Exercise 8.4.11 Let $M_R \in \mathscr{M}od\text{-}R$, $_RN \in R\text{-}\mathscr{M}od$. Show that $\mathrm{id}_M \otimes \mathrm{id}_N = \mathrm{id}_{M \otimes_R N}$.

Exercise 8.4.12 Suppose $M_R, M_R', M_R'' \in \mathscr{M}od\text{-}R$ and $_RN, _RN', _RN'' \in R\text{-}\mathscr{M}od$. Let $f \in \mathrm{Hom}_R(M_R, M_R')$, $f' \in \mathrm{Hom}_R(M_R', M_R'')$ and let $g \in \mathrm{Hom}_R(_RN, _RN')$, $g' \in \mathrm{Hom}_R(_RN', _RN'')$. Show that

$$(f' \circ f) \otimes (g' \circ g) = (f' \otimes g') \circ (f \otimes g).$$

Exercise 8.4.13 Writing $\mathbb{Z}_2$ for $\mathbb{Z}/2\mathbb{Z}$ consider the exact sequence of $\mathbb{Z}$-modules

$$0 \longrightarrow \mathbb{Z} \overset{f}{\longrightarrow} \mathbb{Z} \overset{g}{\longrightarrow} \mathbb{Z}_2 \longrightarrow 0$$

where $f(k) = 2k$. Tensoring with $\mathbb{Z}_2$ yields

$$0 \longrightarrow \mathbb{Z} \otimes \mathbb{Z}_2 \overset{f \otimes \mathrm{id}}{\longrightarrow} \mathbb{Z} \otimes \mathbb{Z}_2 \overset{g \otimes \mathrm{id}}{\longrightarrow} \mathbb{Z}_2 \otimes \mathbb{Z}_2 \longrightarrow 0.$$

Show that $f \otimes \mathrm{id} = 0$ and thus cannot be a monomorphism since $\mathbb{Z} \otimes \mathbb{Z}_2 \cong \mathbb{Z}_2$ by Proposition 8.1.7.

Exercise 8.4.14 Given the short exact sequence in $R\text{-}\mathscr{M}od$

$$0 \longrightarrow {}_RN_1 \overset{f}{\longrightarrow} {}_RN_2 \overset{g}{\longrightarrow} {}_RN_3 \longrightarrow 0$$

with both $_RN_1$ and $_RN_3$ flat, show that the module $_RN_2$ is flat.

Exercise 8.4.15 Let R be a unital ring. Show that, if $\{M_i \mid i \in I\}$ is a family of modules in $\mathscr{M}od\text{-}R$ then $\bigoplus_{i \in I} M_i$ is flat if and only if every M_i is flat.

Exercise 8.4.16 Suppose the functor G: $\mathscr{D} \to \mathscr{C}$ has a left adjoint. Show that G preserves equalizers and hence kernels. (Compare Exercise 4.3.11.)

Exercise 8.4.17 A full subcategory $\mathscr{C}$ of a category $\mathscr{D}$ is called *reflective* if the inclusion functor $\mathscr{C} \hookrightarrow \mathscr{D}$ has a left adjoint, called the *reflector*. Let $\mathscr{M}etric$ be the category of metric spaces (as objects) and uniformly continuous mappings (as morphisms). Show that the category $\mathscr{C}\mathscr{M}etric$ of complete metric spaces is a reflective subcategory and the reflector is the "completion of a metric space".

Exchange Modules and Exchange Rings

9

Our penultimate chapter is devoted to an extension of the concept of semisimple module and, as a consequence, the notion of a semisimple ring. The exchange property for modules was introduced by Crawley and Jónsson in 1964 and the concept of an exchange ring belongs to Warfield (1972). Our main sources for this chapter are [13] and [16], which contain references to the original articles and a lot of additional information.

Throughout this chapter, all rings will be unital, all modules will be unital left modules, and we shall drop the explicit reference to the ring; that is, instead of $_R M$ we simply write M for a unital left module over the ring R.

Definition 9.1 Let R be a unital ring. The module $M \in R\text{-}\mathcal{M}od$ has the *exchange property* if, for every module $A \in R\text{-}\mathcal{M}od$ and every decomposition

$$A = M' \oplus N = \bigoplus_{i \in I} A_i \qquad (9.1.1)$$

with $N \in R\text{-}\mathcal{M}od$, $A_i \in R\text{-}\mathcal{M}od$ and $M' \cong M$, there exist submodules $A_i' \leq A_i$ such that

$$A = M' \oplus \bigoplus_{i \in I} A_i'. \qquad (9.1.2)$$

If the above condition holds for every finite index set I, M is said to have the *finite exchange property*.

The exchange property has a forerunner in the following "Steinitz' Exchange Lemma for semisimple modules":

© The Author(s), under exclusive license to Springer Nature Switzerland AG 2022
M. Mathieu, *Classically Semisimple Rings*,
https://doi.org/10.1007/978-3-031-14209-3_9

Let $A \in R\text{-}\mathcal{M}od$ be semisimple and decomposed as a direct sum $A = \bigoplus_{i \in I} A_i$ of simple submodules $A_i \leq A$. For each submodule $M \leq A$, there is a subset $J \subseteq I$ such that $A = M \oplus \bigoplus_{j \in J} A_j$.

This follows immediately from the arguments in the proof of Proposition 6.1.8.

9.1   Basic Properties of Exchange Modules

We first observe that, the submodules $A_i' \leq A_i$ in Definition 9.1 above are in fact direct summands of the A_i. This follows by applying Exercise 9.4.1 to the chain of inclusions

$$A_i' \subseteq A_i \subseteq A_i' \oplus \left(M' \oplus \bigoplus_{j \neq i} A_i' \right)$$

which yields $A_i = A_i' \oplus C_i$ with $C_i = A_i \cap \left(M' \oplus \bigoplus_{j \neq i} A_i' \right)$ for each $i \in I$.

For finitely generated modules, the finite exchange property already implies the full exchange property.

Proposition 9.1.1 *Let $M \in R\text{-}mod$ be a finitely generated module. If M has the finite exchange property then M has the exchange property.*

Proof Suppose that $A = M' \oplus N = \bigoplus_{i \in I} A_i$; then, as $M' \cong M$, M' is finitely generated and therefore, $M' \subseteq \bigoplus_{j \in J} A_j$ for a finite subset J of I. Applying Exercise 9.4.1 we find that

$$\bigoplus_{j \in J} A_j = M' \oplus \left(N \cap \bigoplus_{j \in J} A_j \right).$$

Since M has the finite exchange property there exist submodules $A_j' \leq A_j$, $j \in J$ such that

$$\bigoplus_{j \in J} A_j = M' \oplus \bigoplus_{j \in J} A_j'$$

and consequently,

$$A = M' \oplus N = \bigoplus_{j \in J} A_j \oplus \bigoplus_{i \in I \setminus J} A_i = M' \oplus \bigoplus_{j \in J} A_j' \oplus \bigoplus_{i \in I \setminus J} A_i,$$

which proves the assertion. $\square$

As it turns out it is in fact sufficient for a module M to have the finite exchange property for two-element decompositions; see Theorem 9.1.5 below. Working towards this statement, we establish the following results.

Proposition 9.1.2 *Let $\{A_i \mid i \in I\}$ be a family in $R\text{-}\mathcal{M}od$ and let $B_i \leq A_i$ for all $i \in I$. Suppose A, M, N and L in $R\text{-}\mathcal{M}od$ are such that*

$$A = M \oplus N \oplus L = \bigoplus_{i \in I} A_i \oplus L,$$

$$A/L = (M + L)/L \oplus \bigoplus_{i \in I} (B_i + L)/L.$$

Then

$$A = M \oplus \bigoplus_{i \in I} B_i \oplus L.$$

Proof The assumption on A/L provides us with $A = M + \sum_{i \in I} B_i + L$. In order to show this is a direct sum, take $m \in M$, $x \in L$ and finitely many $b_i \in B_i$ such that $m + \sum_{i \in I} b_i + x = 0$. Then $m + L + \sum_{i \in I}(b_i + L) = 0$ in A/L wherefore, by assumption, $m \in L$ and each $b_i \in L$. Since the direct sum decomposition of A yields $m = 0$ and $b_i = 0$ (as $B_i \cap L \subseteq A_i \cap L$), it follows that $x = 0$ which was to prove.   □

We say that a module $M \in R\text{-}\mathcal{M}od$ has the c-*exchange property* if the defining property in Definition 9.1 holds for all index sets I of cardinality at most c.

Corollary 9.1.3 *Let $\{A_i \mid i \subset I\}$ be a family in $R\text{-}\mathcal{M}od$ and suppose A, M, N and L in $R\text{-}\mathcal{M}od$ are such that*

$$A = M \oplus N \oplus L = \bigoplus_{i \in I} A_i \oplus L.$$

If M has the c-exchange property then there exist submodules $B_i \leq A_i$, $i \in I$ such that

$$A = M \oplus \bigoplus_{i \in I} B_i \oplus L.$$

Proof Apply Proposition 9.1.2 to $A/L \cong M \oplus L$ and the c-exchange property of M.
                                                                          □

This result will now be used to show that the exchange property is well behaved under direct sums.

Theorem 9.1.4 *Let $M = \bigoplus_{i=1}^{n} M_i$ be a finite direct sum of modules $M_i \in R\text{-}\mathcal{M}od$. Then M has the c-exchange property if and only if each M_i, $1 \leq i \leq n$ has the c-exchange property.*

Remark This result does not extend to infinite direct sums; an example can be found in [13, Example 12.15].

Proof Clearly it suffices to deal with the case $n = 2$; thus assume that $M = X \oplus Y$ for some $X, Y \in R\text{-}\mathcal{M}od$. Suppose that $A = M \oplus N = \bigoplus_{i \in I} A_i$, where I has cardinality at most $\mathfrak{c}$. If Y has the $\mathfrak{c}$-exchange property then, for some $B_i \leq A_i$,

$$A = X \oplus Y \oplus N = Y \oplus \bigoplus_{i \in I} B_i.$$

If X has the $\mathfrak{c}$-exchange property too, then, by Corollary 9.1.3, $A = X \oplus Y \oplus \bigoplus_{i \in I} C_i$ for some submodules $C_i \leq B_i$. It follows that M has the $\mathfrak{c}$-exchange property.

Conversely suppose that $M = X \oplus Y$ has the $\mathfrak{c}$-exchange property. If

$$A = X' \oplus N = \bigoplus_{i \in I} A_i$$

with $X' \cong X$ and $|I| \leq \mathfrak{c}$ then, setting $B = A \oplus Y$, we have

$$B = M' \oplus N = Y \oplus \bigoplus_{i \in I} A_i,$$

where $M' = X' \oplus Y \cong M$. Fix $j \in I$. Since

$$B = M' \oplus N = Y \oplus A_j \oplus \bigoplus_{i \neq j} A_i$$

and M has the $\mathfrak{c}$-exchange property, there exist submodules $C \leq Y \oplus A_j$ and $B_i \leq A_i, i \neq j$, with the property

$$B = M' \oplus C \oplus \bigoplus_{i \neq j} B_i.$$

Applying Exercise 9.4.1 to the inclusions $Y \subseteq Y \oplus C \subseteq Y \oplus A_j$ gives $Y \oplus C = Y \oplus B_j$, where $B_j = (Y \oplus C) \cap A_j$. Hence, $M' \oplus C = (X' \oplus Y) \oplus C = X' \oplus Y \oplus B_j$. When substituting this into the above identity we find

$$B = X' \oplus Y \oplus B_j \oplus \bigoplus_{i \neq j} B_i = X' \oplus Y \oplus \bigoplus_{i \in I} B_i.$$

Since $X' \oplus \bigoplus_{i \in I} B_i \subseteq A$, the modular law (Exercise 2.7.8) entails

$$A \cap \left(Y + \left(X' \oplus \bigoplus_{i \in I} B_i \right) \right) = (A \cap Y) + \left(X' \oplus \bigoplus_{i \in I} B_i \right)$$

and thus, $A = X' \oplus \bigoplus_{i \in I} B_i$ so that X has the c-exchange property. A similar argument yields Y has the c-exchange property too, so the proof is complete. $\square$

Our next result may be somewhat surprising.

Theorem 9.1.5 *A module which has the 2-exchange property has the finite exchange property.*

Proof Suppose M has the 2-exchange property; we shall show that M has the n-exchange property for every $n \in \mathbb{N}$. Let $n > 2$ and assume that M has the $(n-1)$-exchange property. Let $N, A_1, \ldots, A_n \in R\text{-}\mathcal{M}od$ be such that $M \oplus N = \bigoplus_{i=1}^{n} A_i$. Setting $A = \bigoplus_{i=1}^{n} A_i$ and $B = \bigoplus_{i=2}^{n} A_i$ we have $A = A_1 \oplus B$. By the 2-exchange property,

$$A = M \oplus A'_1 \oplus B',$$

where

$$A_1 = A'_1 \oplus A''_1, \quad B = B' \oplus B'' \text{ and } B'' = B \cap (M \oplus A'_1).$$

By the induction hypothesis, since $M \cong A''_1 \oplus B''$, we infer from Theorem 9.1.4 that B'' has the $(n-1)$-exchange property so, from

$$B = B' \oplus B'' = \bigoplus_{i=2}^{n} A_i,$$

we obtain $B = B'' \oplus \bigoplus_{i=2}^{n} A'_i$ for some submodules $A'_i \leq A_i$, $2 \leq i \leq n$. As $B'' = B \cap (M \oplus A'_1)$ and thus $B'' \subseteq M \oplus A'_1 \subseteq B'' \oplus (B' \oplus A_1)$, we can apply Exercise 9.4.1 to find

$$M \oplus A'_1 = B'' \oplus B''' \quad \text{where} \quad B''' = (M \oplus A'_1) \cap (B' \oplus A_1).$$

As a result,

$$A = M \oplus A'_1 \oplus B' = B''' \oplus B'' \oplus B'$$

$$= B \oplus B''' = \bigoplus_{i=2}^{n} A'_i \oplus B'' \oplus B'''$$

$$= M \oplus \bigoplus_{i=1}^{n} A'_i$$

entailing that M has the n-exchange property. $\square$

Corollary 9.1.6 *Let $M \in R$-mod. If M has the 2-exchange property then it has the exchange property.*

Proof This follows immediately from Proposition 9.1.1 and Theorem 9.1.5.   □

It may be time to see a module which fails the (2-)exchange property.

Example 9.1.7 The module $\mathbb{Z}$ fails the 2-exchange property. Note at first that $\mathbb{Z}$ is indecomposable as a module: if $\mathbb{Z} \cong X \oplus Y$ with $X, Y \in \mathbb{Z}$-*Mod* then X or Y must be trivial. Now start with the identity

$$\mathbb{Z} \oplus \mathbb{Z} = \mathbb{Z}(1, 0) \oplus \mathbb{Z}(0, 1) = \mathbb{Z}(7, 3) \oplus \mathbb{Z}(5, 2)$$

and suppose that $\mathbb{Z}$ had the 2-exchange property. Then there are submodules $X \leq \mathbb{Z}(7, 3), Y \leq \mathbb{Z}(5, 2)$ such that $\mathbb{Z} \oplus \mathbb{Z} = \mathbb{Z}(1, 0) \oplus X \oplus Y$. It follows that $\mathbb{Z} \cong X \oplus Y$ whence $X = 0$ or $Y = 0$. Therefore

$$\mathbb{Z} \oplus \mathbb{Z} = \mathbb{Z}(1, 0) \oplus \mathbb{Z}(7, 3) \quad \text{or} \quad \mathbb{Z} \oplus \mathbb{Z} = \mathbb{Z}(1, 0) \oplus \mathbb{Z}(5, 2)$$

which is impossible as neither $\{(1, 0), (7, 3)\}$ nor $\{(1, 0), (5, 2)\}$ form a $\mathbb{Z}$-basis of $\mathbb{Z} \oplus \mathbb{Z}$.

In fact, indecomposable modules which have the exchange property can be characterised as those with local endomorphism ring [16, Theorem 2.8].

9.2   Exchange Rings

Definition 9.2.1 A unital ring R is said to be an *exchange ring* if, for any pair of elements a and b in R with $a + b = 1$, there exist idempotents $e \in Ra$ and $f \in Rb$ such that $e + f = 1$.

In this section, we will show that R is an exchange ring if and only if the standard left module $_RR$ has the exchange property and study some of the properties of this type of ring. It appears that we have defined a property that depends on the choice of the standard left module $_RR$ over its right-handed analogue R_R; however Proposition 9.2.3 below shows that this is not the case.

We prepare this by the following lemma.

Lemma 9.2.2 *The following conditions are equivalent for an element x in a unital ring R.*

(a) $\exists\, e = e^2 \in R : e - x \in R(x - x^2)$;

(b) $\exists\, e = e^2 \in Rx, c \in R : (1 - e) - c(1 - x) \in \text{rad}(R)$;

(c) $\exists\, e = e^2 \in Rx: R = Re + R(1 - x)$;

(d) $\exists\, e = e^2 \in Rx: 1 - e \in R(1 - x)$.

Proof (a) $\Rightarrow$ (b) Suppose $e - x = r(x - x^2)$ where e is an idempotent in R and $r \in R$. Then $e = r(1 - x)x + x \in Rx$ and

$$1 - e - (1 - rx)(1 - x) = 0 \in \mathrm{rad}(R).$$

(b) $\Rightarrow$ (c) If $1 - (e + c(1 - x)) \in \mathrm{rad}(R)$ for some $c \in R$ then $e + c(1 - x)$ is invertible in R and therefore, $R = Re + R(1 - x)$.

(c) $\Rightarrow$ (d) Let $e \in Rx$ be an idempotent such that, for some $s, t \in R$, $1 = te + s(1 - x)$. Put $f = e + (1 - e)te$; then f is an idempotent in Rx and

$$1 - f = 1 - e - (1 - e)te = (1 - e)s(1 - x) \in R(1 - x).$$

(d) $\Rightarrow$ (a) If $1 - e \in R(1 - x)$ for some idempotent $e \in Rx$ then $e - x = e(1 - x) - (1 - e)x \in R(x - x^2)$. $\qquad\qquad\square$

Proposition 9.2.3 *A unital ring R is an exchange ring if any, and hence every, of the conditions in Lemma 9.2.2 holds for all $x \in R$. Moreover, this is equivalent to the condition that, for any pair a and b in R with $a + b = 1$, there exist idempotents $e \in aR$ and $f \in bR$ such that $e + f = 1$.*

Proof Since it is evident that condition (d) in Lemma 9.2.2 is equivalent to the condition that, for any pair of elements a and b in R with $a + b = 1$, there exist idempotents $e \in Ra$ and $f \in Rb$ such that $e + f = 1$, the first assertion in the proposition holds.

Now assume that the condition in the second assertion of the proposition is satisfied; we show that this implies that R is an exchange ring, and the converse will be true by symmetry. Take $a, b \in R$ with $a + b = 1$ and pick $e = e^2 \in aR$, $f = f^2 \in bR$ such that $e + f = 1$. Choose $r \in Re$, $s \in Rf$ with the property $e = ar$ and $f = bs$. It follows that $r = re = rar$ and $rbs = rebs = ref = 0$ and similarly, $s = sfs$ and $sar = 0$. Put

$$r' = 1 - sb + rb \quad \text{and} \quad s' = 1 - ra + sa.$$

Then $r's = 0 = s'r$ and, as $ar + bs = e + f = 1$, we find

$$ar' = a(1 - sb) + arb = a(1 - sb) + (1 - bs)b = a(1 - bs) + b(1 - sb) = 1 - sb.$$

This entails $r'ar' = r'(1 - sb) = r'$, and a similar argument yields $bs' = 1 - ra$ and $s'bs' = s'$. Put $e' = r'a \in Ra$ and $f' = s'b \in Rb$ to obtain idempotents which add up to 1:

$$e' + f' = r'a + s'B = (1 - sb + rb)a + (1 - ra + sa)b$$
$$= a - sba + rba + b - rab + sab$$
$$= a + b + s(ab - ba) + r(ba - ab) = a + b = 1$$

since a and $b = 1 - a$ commute. This was to prove.   □

Corollary 9.2.4 *Every homomorphic image of an exchange ring is an exchange ring.*

Proof This follows immediately from the definition or any of the conditions listed in Lemma 9.2.2.   □

Let L be a subgroup of $(R, +)$. One says that *idempotents lift modulo L* if, whenever $x \in R$ satisfies $x - x^2 \in L$, there is an idempotent $e \in R$ such that $e - x \in L$. Proposition 9.2.3 entails the following characterisation of exchange rings.

Corollary 9.2.5 *A unital ring R is an exchange ring if and only if idempotents lift modulo every left ideal of R.*

We will now make contact with the exchange property, which will justify the choice of the terminology 'exchange ring'.

Theorem 9.2.6 *Let R be a unital ring. Then $M \in R\text{-}\mathcal{M}od$ has the finite exchange property if and only if $\text{End}_R(M)$ is an exchange ring. In particular, R is an exchange ring if and only if $_R R$ has the exchange property.*

Proof By Proposition 9.2.3, R is an exchange ring if and only if its opposite ring R^{op} is an exchange ring. Since $R^{op} \cong \text{End}_R(_R R)$, see Exercise 2.7.1, the statement on R follows immediately from the main statement in the theorem together with Corollary 9.1.6.

Put $S = \text{End}_R(M)$ and suppose S is an exchange ring. In order to deduce that M has the finite exchange property, it suffices to show that M has the 2-exchange property (Theorem 9.1.5). Assume, thus, that

$$A = M \oplus N = A_1 \oplus A_2 \quad \text{in } R\text{-}\mathcal{M}od. \tag{9.2.1}$$

Denote by $p \in S$ the projection onto M and by $q_i \in S$ the projections onto A_i, $i = 1, 2$, respectively. These are idempotents satisfying $p = pq_1 p + pq_2 p$. Since $pSp \cong \text{End}_R(M)$ as rings, by hypothesis, pSp is an exchange ring. We can therefore apply

Lemma 9.2.2 to pSp to choose orthogonal idempotents $s_i \in pSp(pq_i p)$, $i = 1, 2$ with $s_1 + s_2 = p$ (use condition (a) in Lemma 9.2.2 and note that $pq_2 p = p - pq_1 p$). We can write $s_i = a_i q_i p$ for some $a_i \in pSp$ with $s_i a_i = a_i$, $i = 1, 2$. Define $t_i \in S$ by $t_i = q_i a_i q_i$, $i = 1, 2$. Then $t_i t_j = 0$ for $i \neq j$ and

$$t_i - t_i^2 = q_i a_i q_i - q_i a_i q_i q_i a_i q_i = q_i a_i p q_i - q_i a_i p q_i a_i p q_i = q_i (s_i - s_i^2) = 0$$

for $i = 1, 2$. Hence, t_1, t_2 are orthogonal idempotents with $t_i A \subseteq A_i$, $i = 1, 2$. As a result, $A_i = t_i A \oplus A_i'$ with $A_i' = A_i \cap \ker t_i$, $i = 1, 2$, and we obtain

$$A = (t_1 A \oplus t_2 A) \oplus A_1' \oplus A_2'.$$

The argument will be finished once we show that $A = M \oplus A_1' \oplus A_2'$.

Note first that $t_i p = q_i a_i q_i p = q_i s_i$ and thus,

$$a_i t_i p = a_i q_i s_i = a_i q_i a_i q_i p = a_i q_i p a_i q_i p = s_i^2 = s_i$$

for $i = 1, 2$. Suppose $x \in M \cap (A_1' \oplus A_2')$. Then $t_1 x = 0 = t_2 x$ and therefore,

$$x = px = s_1 x + s_2 x = a_1 t_1 px + a_2 t_2 px = a_1 t_1 x + a_2 t_2 x = 0.$$

It follows that $M \cap (A_1' \oplus A_2') = 0$.

Next take $x \in A$ and write it (uniquely) as $x = x_1 + x_2 + y$ with $x_i \in t_i A$, $i = 1, 2$, and $y \in A_1' \oplus A_2'$. Observe that

$$t_i a_j t_j = t_i p a_j q_j t_j = q_i s_i s_j a_j q_j t_j = \delta_{ij} t_j$$

for $i, j = 1, 2$. It follows that $x_i - a_i x_i \in A_i'$ for each i and consequently,

$$x - a_1 x_1 - a_2 x_2 = x_1 - a_1 x_1 + x_2 - a_2 x_2 + y \in A_1' \oplus A_2'.$$

Since $a_i x_i \in M$ for each i this shows that $A = M \oplus A_1' \oplus A_2'$ proving that M has the 2-exchange property.

In order to prove the "only if"-part suppose that M has the finite exchange property. Let $a, b \in S = \operatorname{End}_R(M)$ be such that $a + b = 1$. Write $A := M \oplus M = N_1 \oplus N_2$, where N_i is the image of the ith canonical projection from A onto M, $i = 1, 2$. Put $D = \{(x, x) \mid x \in M\}$ and $M' = \{(ax, -bx) \mid x \in M\}$. Then $M' \cong M$ and $A = M' \oplus D = N_1 \oplus N_2$. By hypothesis, there exist $N_i' \leq N_i$, $i = 1, 2$ with the property $A = M \oplus N_1' \oplus N_2'$. Consequently, for each $x \in M$, there is a unique decomposition

$$(x, x) = (ay, -by) + (x_1, 0) + (0, x_2) \tag{9.2.2}$$

with $y \in M$ and $(x_1, 0) \in N_1'$, $(0, x_2) \in N_2'$. We define $a', b' \in S$ by $a'x = x_2$ and $b'x = x_1$. Then

$$(aa'x, aa'x) = (aa'x, -ba'x) + (0, 0) + (0, a'x),$$

$$(bb'x, bb'x) = (-ab'x, bb'x) + (b'x, 0) + (0, 0)$$

and thus, $a'aa' = a'$ and $b'bb' = b'$ so that $e = aa' \in aS$ and $f = bb' \in bS$ are idempotents. From (9.2.2) we obtain $x = ay + b'x = -by + a'x$ and therefore, $y = (a + b)y = (a' - b')x$. Substituting back we find

$$x = a(a' - b')x + b'x = aa'x + (1 - a)b'x = aa'x + bb'x$$

which implies that $e + f = 1$. This proves that $\text{End}_R (M)$ is an exchange ring, in view of Proposition 9.2.3. $\square$

Remark The concept of an exchange ring was originally introduced by Warfield but the equivalent formulation via Definition 9.2.1 is due to Nicholson; the proof of Theorem 9.2.6 presented here is the original one in [28].

Corollary 9.2.7 *A unital ring R is an exchange ring if and only if every finitely generated projective module in R-$\mathcal{M}od$ has the exchange property.*

Proof Only the "only if"-part needs proof. By Theorem 9.2.6, $_R R$ has the exchange property and therefore, every finitely generated free R-module has the exchange property, by Theorem 9.1.4. As every finitely generated projective module is a direct summand of a finitely generated free module, the same theorem yields the claim.
                                                                                    $\square$

Another characterisation of exchange rings involves the Jacobson radical.

Theorem 9.2.8 *A unital ring R is an exchange ring if and only if the quotient ring $R/\text{rad}(R)$ is an exchange ring and idempotents lift modulo $\text{rad}(R)$.*

Proof The "only if"-statement follows from Corollaries 9.2.4 and 9.2.5. Now suppose that $R' = R/\text{rad}(R)$ is an exchange ring and let $x \in R$. Write x' for $x + \text{rad}(R)$. There are an idempotent $a' \in R'x'$ and an element $c' \in R'$ such that $1 - a' = c'(1 - x')$, by condition (b) in Lemma 9.2.2. Without loss of generality we can assume that $a \in Rx$. If idempotents lift modulo $\text{rad}(R)$, there is $f = f^2 \in R$ such that $f' = a'$. Set $u = 1 - f + a$, which is a unit in R. It follows that $e = u^{-1} f u = u^{-1} f a$ is an idempotent in Rx. Moreover $e' = f' = a'$ and thus $(1 - e) - c(1 - x) \in \text{rad}(R)$ which proves that R is an exchange ring by Lemma 9.2.2.
                                                                                    $\square$

Here is a list of some examples of exchange rings.

Examples 9.2.9

1. Every von Neumann regular ring R is an exchange ring (cf. Exercise 7.4.13). Let $x \in R$ and take $y \in R$ with $xyx = x$. Then $f = yx$ is an idempotent in Rx. Put $e = f + (1 - f)xf$ to obtain an idempotent $e \in Rx$ such that $1 - e = (1 - f)(1 - xf) \in R(1 - x)$. The statement follows from Lemma 9.2.2.
2. Call a unital ring R *clean* if every element in R is the sum of a unit and an idempotent. Every clean ring is an exchange ring. For, take $x \in R$ and write it as $x = u + f$ where u is a unit and f is an idempotent. Set $e = u^{-1}(1 - f)u$. Then

$$u(e - x) = (1 - f)u - u(u + f) = u - fu - uf - u^2 = x - x^2$$

and thus, $e - x \in R(x - x^2)$ so that Lemma 9.2.2 and Proposition 9.2.3 yield the claim.

Remark 9.2.10 Exchange rings have rather interesting properties. For instance, it was shown by Ara et al. [4] that a unital C^*-algebra is an exchange ring if and only if it has real rank zero, a rather important structural property useful for the classification of C^*-algebras. Ara [3] also introduced the concept of *non-unital* exchange rings proving, amongst others, that a (possibly non-unital) ring R is an exchange ring if and only if it contains an ideal I such that both I and R/I are exchange rings and idempotents lift modulo I. For these, and further comments, see [13, 11.43].

9.3   Commutative Exchange Rings

In this section we shall study some special properties of commutative exchange rings. We have already observed (Exercise 9.4.5) that a commutative (unital) ring is a exchange ring if and only if it is clean. We will now establish a connection with various kinds of dimension for a commutative ring.

A good proportion of the interest in commutative rings comes from their appearance in Analysis, especially as rings of continuous functions. We will assume the reader has a basic knowledge in (point-set) Topology, such as provided by, e.g., [32]. For a compact Hausdorff space X, we denote by $C_{\mathbb{R}}(X)$ the space of continuous real-valued functions on X, which is a commutative unital ring under pointwise operations. (Compare also Example 3.2.2)

The Lebesgue covering dimension of X is defined via refinements of open coverings of X; see [19], for example. However, we will only need the case when X is *zero-dimensional* which is equivalent to the condition that X is totally disconnected, that is, singleton sets are the only non-empty connected subsets of X. This in turn is equivalent to the existence of a basis for the topology of X consisting of clopen (= closed and open) subsets of X. We will use this latter characterisation since it entails that the ring $C_{\mathbb{R}}(X)$ is abundant with idempotents: for each clopen subset $U \subseteq X$, its characteristic function χ_U belongs to $C_{\mathbb{R}}(X)$ and is an idempotent.

The following is a standard characterisation of zero-dimensional compact Hausdorff spaces; see, e.g., [19, Theorem 16.17].

Proposition 9.3.1 *The following conditions on a compact Hausdorff space X are equivalent.*

(a) *X is zero-dimensional;*
(b) *for every pair of disjoint closed subsets A and B of X, there exist disjoint open subsets U and V of X such that $X = U \cup V$, $A \subseteq U$ and $B \subseteq V$;*
(c) *for every closed subset A of X and every open subset U of X containing A, there exists a clopen subset V of X such that $A \subseteq V \subseteq U$.*

The exchange property for $C_{\mathbb{R}}(X)$ turns out to be the equivalent of zero-dimensionality of X.

Proposition 9.3.2 *A compact Hausdorff space X is zero-dimensional if and only if the ring $C_{\mathbb{R}}(X)$ is an exchange ring.*

Proof Suppose at first that X is zero-dimensional. Let $a, b \in C_{\mathbb{R}}(X)$ satisfy $a + b = 1$ and let $A = a^{-1}(0)$ and $B = b^{-1}(0)$. Then A and B are disjoint closed subsets of X so, by Proposition 9.3.1, there exist $U, V \subseteq X$ clopen and disjoint such that $X = U \cup V$, $A \subseteq U$ and $B \subseteq V$. Set $e = \chi_V$ and $f = \chi_U$, which are idempotents in $C_{\mathbb{R}}(X)$ with sum 1. Define $c \in C_{\mathbb{R}}(X)$ by $c = \frac{1}{a}e$ and $d \in C_{\mathbb{R}}(X)$ by $d = \frac{1}{b}f$; more precisely, $c(t) = \begin{cases} \frac{1}{a(t)} & \text{if } t \in V \\ 0 & \text{if } t \in U \end{cases}$ and similarly for d. Then $e = ca$ and $f = db$ wherefore, by Definition 9.2.1, $C_{\mathbb{R}}(X)$ is an exchange ring.

Now suppose that $C_{\mathbb{R}}(X)$ is an exchange ring. Let A and B be two disjoint closed subsets of X. By Urysohn's Lemma (see, e.g., Theorem 1.5.6 together with Theorem 1.6.6 in [32]) there is $a \in C_{\mathbb{R}}(X)$ such that $a(X) \subseteq [0, 1]$, $A \subseteq a^{-1}(0)$ and $B \subseteq a^{-1}(1)$. Let $e \in C_{\mathbb{R}}(X)$ be an idempotent such that $e = ca$ and $1 - e = d(1 - a)$ for some $c, d \in C_{\mathbb{R}}(X)$. Let $V \subseteq X$ be such that $e = \chi_V$. Then V is clopen and, with $U = X \setminus V$, $A \subseteq U$ and $B \subseteq V$. By Proposition 9.3.1, X is zero-dimensional. $\square$

Next we introduce two algebraic versions of dimension for a general commutative unital ring that appeared in the work of Canfell ([11] and [12]). For the remainder of this section, R will always denote a commutative unital ring and for an element $a \in R$, the ideal generated by a will be written as (a).

Definition 9.3.3 A finite set $(a_1), \ldots, (a_n)$ of principal ideals of R is said to be *uniquely generated* if, whenever $(a_i) = (b_i)$, $1 \le i \le n$ for a set $\{b_1, \ldots, b_n\} \subseteq R$, there exist elements $u_i \in R$ such that $a_i = b_i u_i$ for all $1 \leq i \leq n$ and $R = (u_1) + \cdots + (u_n)$. The *first Canfell dimension* of R, denoted by $\dim_1 R$, is the least natural number n such that every set of $n + 1$ principal ideals is uniquely generated.

Definition 9.3.4 The *order* of a finite subset $\{a_1, \ldots, a_n\} \subseteq R$ which generates R is the largest $m \in \{1, \ldots, n\}$ with the property that $m + 1$ members of $\{a_1, \ldots, a_n\}$ have non-zero product. A *refinement* of $\{a_1, \ldots, a_n\}$ is another finite subset $\{b_1, \ldots, b_k\} \subseteq R$ such that, for each $1 \leq j \leq k$, there is $1 \leq i \leq n$ with $b_j \in (a_i)$. The *second Canfell dimension* of R, denoted by $\dim_2 R$, is the least $d \in \mathbb{N}$ such that every generating subset of R has a refinement of order at most d.

Both Canfell dimensions of $C_{\mathbb{R}}(X)$ agree with the covering dimension of the compact Hausdorff space X, which was shown in [11] and [12], respectively. Consequently, by Proposition 9.3.2, $C_{\mathbb{R}}(X)$ is an exchange ring if and only if $\dim_1 C_{\mathbb{R}}(X) = 0$ which is equivalent to $\dim_2 C_{\mathbb{R}}(X) = 0$. These results partly extend to arbitrary commutative rings.

Theorem 9.3.5 *Let R be a commutative unital ring. If R is an exchange ring then $\dim_1 R = 0$.*

Proof Let $a, b \in R$ be such that $(a) = (b)$. Then there are $s, t \in R$ with $a = bs$ and $b = at$ so that $b(1 - st) = 0$. Since R is an exchange ring, there is an idempotent $e \in R$ such that $e \in (1 - st)R$ and $1 - e \in stR$. For some $v, u \in R$ we thus have $e = (1 - st)v$ and $1 - e = stu$. Replacing v by ve and u by $u(1 - e)$ if necessary, we can assume that $v = ve$ and $u = u(1 - e)$. Note that $be = b(1 - st)v = 0$ so that $b = b(1 - e)$.

Let $k = s(1 - e) + (1 - st)e$. Then k is invertible with inverse $tu + v$:

$$k(tu + v) = \big(s(1 - e) + (1 - st)e\big)\big(tu(1 - e) + ve\big)$$

$$= stu(1 - e) + (1 - st)ve = 1 - e + e = 1.$$

In addition,

$$bk = b(s(1 - e) + (1 - st)e) = b(1 - e)s + b(1 - st)e = bs = a$$

wherefore the first Canfell dimension of R must be zero. $\square$

Remark 9.3.6 Every integral domain has first Canfell dimension zero. This follows immediately from $a = bs$ and $b = at$, hence $b(1 - st) = 0$ so that $st = 1$ (unless $b = 0$, the trivial case). However, not every integral domain is an exchange ring; e.g., $\mathbb{Z}$ is not (Example 9.1.7).

The relation to the second Canfell dimension is even more pleasing.

Theorem 9.3.7 *Let R be a commutative unital ring. Then R is an exchange ring if and only if* $\dim_2 R = 0$.

Proof Let us first observe that $\dim_2 R = 0$ means that, whenever $a_1, \ldots, a_n \in R$ with $(a_1) + \cdots + (a_n) = R$ are given, there exist $b_1, \ldots, b_m \in R$ with $(b_1) + \cdots + (b_m) = R$ and, for each $1 \leq i \leq m$, there is $1 \leq k \leq n$ with $b_i \in (a_k)$ with the property that $b_i b_j = 0$ for all $i, j, i \neq j$.

Now suppose that R is an exchange ring. Let $a_1, \ldots, a_n \in R$ and $r_1, \ldots, r_n \in R$ be such that $\sum_{i=1}^n a_i r_i = 1$. By Exercise 9.4.6, there exist orthogonal idempotents $e_i \in Ra_i r_i$, $1 \leq i \leq n$ such that $\sum_{i=1}^n e_i = 1$. Since $e_i \in (a_i)$ for all i it follows that $\dim_2 R = 0$.

Suppose $\dim_2 R = 0$. Take $a \in R$; as $R = aR + (1-a)R$, by hypothesis, there exist $a_1, \ldots, a_\ell \in aR$ and $a_{\ell+1}, \ldots, a_m \in (1-a)R$ such that $a_i a_j = 0$ for $i \neq j$ and

$$a_1 r_1 + \cdots + a_m r_m = 1$$

for some $r_i \in R$, $1 \leq i \leq m$. Letting $e = a_1 r_1 + \cdots + a_\ell r_\ell$ we have $e \in aR$ and $1 - e \in (1-a)R$. Since

$$e(1-e) = (a_1 r_1 + \cdots + a_\ell r_\ell)(a_{\ell+1} r_{\ell+1} + \cdots + a_m r_m) = 0,$$

as all the a_is are mutually orthogonal, we find $e = e^2$ and R is an exchange ring. $\square$

In the remainder of this section we will introduce and investigate a variety of concepts for commutative unital rings that all turn out to be closely related to the notion of an exchange ring. Our source for this is a paper by McGovern [27], which also contains the references to the original papers where these concepts were first put forward.

Definition 9.3.8 Let R be a commutative unital ring. Then

(i) R is a *Gelfand ring* if, for every pair of elements a and b in R with $a + b = 1$, there exist $r, s \in R$ such that $(1 + ar)(1 + bs) = 0$.

(ii) R is a *pm-ring* if every prime ideal of R is contained in a unique maximal ideal.

(iii) R is a *tb-ring* if, for every pair of distinct maximal ideals of R, there is an idempotent belonging to exactly one of them.

There is a canonical topological space associated to every commutative unital ring.

Definition 9.3.9 For a commutative unital ring R, let $\operatorname{Spec}(R)$ denote the set of all prime ideals of R endowed with the following topology: a basis for the open subsets is given by sets of the form

$$U(a) = \{P \in \operatorname{Spec}(R) \mid a \notin P\},$$

$a \in R$. This is the *prime ideal space* or *spectrum* of R and the topology is referred to as the *hull-kernel topology* or *Zariski topology*.

The prime ideal property implies that $U(ab) = U(a) \cap U(b)$ for all $a, b \in R$. It is customary to denote the complement of $U(a)$ by $V(a)$, that is, $V(a) = \{P \in \operatorname{Spec}(R) \mid a \in P\}$. In general, $\operatorname{Spec}(R)$ has poor separation properties; its closed points are the maximal ideals in R. The subset $\operatorname{Max}(R)$ of all maximal ideals is a subspace of $\operatorname{Spec}(R)$ when equipped with the subspace topology, and is called the *maximal spectrum* of R. The basic subsets in $\operatorname{Max}(R)$ are thus $U'(a) = U(a) \cap \operatorname{Max}(R)$, $a \in R$, and their complements will likewise be denoted by $V'(a)$. Nevertheless both $\operatorname{Spec}(R)$ and $\operatorname{Max}(R)$ are compact spaces. More details on the prime spectrum can be found in [6] or [10, Chap. II].

Notation Let $E(R)$ denote the following subset of $\operatorname{Max}(R)$:

$$E(R) = \{U'(e) \mid e = e^2 \in R\}.$$

The next result provides us with the relations between the various types of rings defined above.

Theorem 9.3.10 *The following conditions on a commutative unital ring R are equivalent.*

(a) *R is an exchange ring;*
(b) *R is a Gelfand ring and $\operatorname{Max}(R)$ is zero-dimensional;*
(c) *R is a pm-ring and $\operatorname{Max}(R)$ is zero-dimensional;*
(d) *R is a tb-ring;*
(e) *$E(R)$ is a basis for the Zariski topology on $\operatorname{Max}(R)$;*
(f) *for each $a \in R$, there is an idempotent $e \in R$ such that $V'(a) \subseteq U'(e)$ and $V'(1-a) \subseteq U'(1-e)$.*

The proof of Theorem 9.3.10 will be split up into a series of propositions which emphasise more clearly the interconnections between the different conditions. We start by examining *pm-rings*; clearly the acronym stems from the assumption that, for each prime ideal, there is a unique maximal ideal containing it.

Proposition 9.3.11 *A commutative unital ring R is a pm-ring if and only if, for every pair of distinct maximal ideals M and M' of R, there exist $a \notin M$ and $b \notin M'$ such that $ab = 0$. In this case, $\mathrm{Max}(R)$ is a Hausdorff space.*

Proof Let P be a prime ideal in R and let M be a maximal ideal containing P. If the condition is satisfied, for every other maximal ideal $M' \neq M$, take $a \notin M$ and $b \notin M'$ such that $ab = 0$. As $ab \in P$ it follows that $b \in P$ and hence $P \nsubseteq M'$. Therefore R is a pm-ring.

Conversely, take two different maximal ideals M, M' of R and put

$$S = \{ab \mid a \notin M, \, b \notin M'\}.$$

If the semigroup S does not contain contain 0, a standard argument using Zorn's Lemma yields the existence of a prime ideal P with $P \cap S = \emptyset$. Hence $P \subseteq M \cap M'$ and R is not a pm-ring.

In order to show that $\mathrm{Max}(R)$ is Hausdorff whenever R is a pm-ring let $M, M' \in \mathrm{Max}(R)$, $M \neq M'$. Take $a \notin M$ and $b \notin M'$ such that $ab = 0$. Then $M \in U'(a)$, $M' \in U'(b)$ and $U'(a) \cap U'(b) = U'(ab) = \emptyset$ which proves the claim. □

Let X be a compact Hausdorff space; then $C_{\mathbb{R}}(X)$ is a pm-ring, see [19, Theorem 2.11 together with Remark 6.6 (c)] or [19, Theorem 7.15]. Since $\mathrm{Max}(C_{\mathbb{R}}(X))$ is homeomorphic to X (e.g., [6, Exercise 26 in Chap. I] or [19, Chap. 7]), $\dim \mathrm{Max}(C_{\mathbb{R}}(X)) = 0$ is equivalent to $C_{\mathbb{R}}(X)$ being an exchange ring (Proposition 9.3.2). Consequently, the equivalence of conditions (a) and (c) in Theorem 9.3.10 is a direct generalisation of Proposition 9.3.2. Working towards this equivalence, we note the following.

Proposition 9.3.12 *Let R be a commutative unital ring. Then*

$$R \text{ exchange ring} \implies R \text{ tb-ring} \implies R \text{ pm-ring}. \tag{9.3.3}$$

Proof Suppose first that R is an exchange ring. Let $M, M' \in \mathrm{Max}(R)$ with $M \neq M'$. Take $a \in M' \setminus M$ and observe that $(a) + M = R$. Thus, for some $x \in M$, $r \in R$, $1 = ar + x$. By hypothesis, there is an idempotent $e \in Rx \subseteq M$ such that $1 - e \in R(1 - x) = Rar \subseteq M'$. It follows that $e \notin M'$ and R is a tb-ring.

Suppose now that R is a tb-ring. Let $M, M' \in \mathrm{Max}(R)$ with $M \neq M'$ and let $e \in M \setminus M'$ be an idempotent. Since $1 - e \notin M$ and $e(1 - e) = 0 \in M'$ it follows that $1 - e \in M'$ entailing that R is a pm-ring, by Proposition 9.3.11. □

Corollary 9.3.13 *Let R be a commutative exchange ring. Then the space $\mathrm{Max}(R)$ is a compact Hausdorff space.*

The terminology "tb-ring" stands for "topologically boolean ring" and is motivated by the fact that a compact zero-dimensional Hausdorff space is also called a *boolean space* and the result below.

Proposition 9.3.14 *Let R be a tb-ring. Then* $\text{Max}(R)$ *is zero-dimensional.*

Proof Note first that $\text{Max}(R)$ is a compact Hausdorff space, by Propositions 9.3.11 and 9.3.12. It therefore suffices to show that the topology on $\text{Max}(R)$ has a basis consisting of clopen subsets. In fact, these will be the elements of $E(R)$. By Exercise 9.4.9, each $U'(e)$, e an idempotent in R, is clopen and, whenever a closed subset $K \subseteq \text{Max}(R)$ is covered by a family $\{U'(e_i) \mid i \in I\}$ of subsets from $E(R)$, by compactness, a finite subfamily suffices to cover K and, by Exercise 9.4.9, their union belongs to $E(R)$, so K is in fact contained in one subset of the form $U'(e)$.

Let $K \subseteq \text{Max}(R)$ be closed and let $M \in \text{Max}(R)$ be in the complement of K. For each $N \in K$, there is an idempotent $e_N \notin N$ which belongs to M. Then $N \in U'(e_N)$, $M \in U'(1 - e_N)$ and $U'(e_N) \cap U'(1 - e_N) = \emptyset$. Since $K \subseteq \bigcup_{N \in K} U'(e_N)$, the above argument yields an idempotent $e \in R$ such that $K \subseteq U'(e)$ and $M \notin U'(e)$. It follows that $E(R)$ is a basis for the Zariski topology on $\text{Max}(R)$. $\square$

Remark 9.3.15 The argument in the above proof in fact establishes the implication (d) $\Rightarrow$ (e) in Theorem 9.3.10. Let us now assume that condition (e) holds. This entails that $E(R)$ is the collection of *all* clopen subsets of $\text{Max}(R)$. Take $a \in R$; then $V'(a)$ is a closed subset of $\text{Max}(R)$ and hence contained in some $U'(e)$, $e = e^2$ by the above argument. Likewise $V'(1 - a) \subseteq U'(1 - e)$ and hence condition (f) in Theorem 9.3.10 holds.

Assuming (f) we will now show that R must be an exchange ring. Take $a \in R$. By hypothesis, there is an idempotent $e \in R$ such that $V'(a) \subseteq U'(e)$ and $V'(1 - a) \subseteq U'(1 - e)$. Let M be a maximal ideal of R. Suppose $a \in M$; then $e \notin M$ and thus $e - a \notin M$. Suppose $a \notin M$; if $e - a \in M$ then $a + M = e + M = 1 + M$ since R/M is a field and $e + M$ is a non-zero idempotent. It follows that $1 - a \in M$ so $1 - e \notin M$ wherefore $e \in M$, which is impossible. As a result, $e - a$ is invertible in R and therefore, $a = e - a + e$ can be written as a sum of a unit and an idempotent, in other words, R is a clean ring. It follows from Example 9.2.9.2 that R is an exchange ring.

In summary we have established so far the following chain of inclusions

$$(a) \Rightarrow (d) \Rightarrow (e) \Rightarrow (f) \Rightarrow (a)$$

of the conditions in Theorem 9.3.10, by Proposition 9.3.12 and Remark 9.3.15, as well as (d) $\Rightarrow$ (c) by Propositions 9.3.12 and 9.3.14. We shall now see that the properties 'Gelfand ring' and 'pm-ring' are always equivalent for a commutative unital ring so that conditions (b) and (c) are equivalent as well.

Proposition 9.3.16 *A commutative unital ring R is a Gelfand ring if and only if it is a pm-ring.*

Proof Suppose R is a Gelfand ring. Let M and M' be two distinct maximal ideals in R. For any $m \in M \setminus M'$, $M' + (m) = R$ and thus, $m' + rm = 1$ for some $m' \in M'$ and $r \in R$. By assumption, there are $s, t \in R$ such that $(1 + sm')(1 + trm) = 0$. Since $1 + sm' \notin M'$ and $1 + trm \notin M$ Proposition 9.3.11 entails that R is a pm-ring.

Now take $a, b \in R$ with $a + b = 1$ and suppose that R fails to be a Gelfand ring. Then the semigroup

$$S = \{(1 + ar)(1 + bs) \mid r, s \in R\}$$

does not contain 0. The usual argument invoking Zorn's Lemma yields the existence of a prime ideal P with $P \cap S = \emptyset$. Since $P + (a) \neq R$ (otherwise $1 = x - ar$ with $x \in P$, $r \in R$ would force $x = 1 + ar \in S$ which is impossible) there is a maximal ideal $M \subset R$ containing $P + (a)$. Similarly, there is a maximal ideal $M' \subset R$ containing $P + (b)$. As $M \neq M'$ (for $a + b = 1$), Proposition 9.3.11 implies that R is not a pm-ring. □

To complete the proof of Theorem 9.3.10, we need to show that one of the conditions (b) or (c) in the theorem imply that R is an exchange ring. The strategy to achieve this is to construct two suitable orthogonal ideals of R whose sum is R. These will arise as intersections of maximal ideals, and since the intersection of maximal ideals is not a maximal but only a prime ideal, we will have to work within the prime spectrum for the first time.

Proposition 9.3.17 *Let R be a pm-ring whose maximal spectrum is zero-dimensional. Then R is an exchange ring.*

Proof We aim to show that $E(R)$ is a basis for the Zariski topology of $\mathrm{Max}(R)$; as seen in Remark 9.3.15 above, this implies that R is an exchange ring. First assume that the nil radical $\mathrm{nil}(R)$ is zero. Let $K \subseteq \mathrm{Max}(R)$ be a clopen subset. Set

$$\overline{K} = \{P \in \mathrm{Spec}(R) \mid P \subseteq M \text{ for some } M \in K\}.$$

As $\overline{K}$ is the preimage of K under the canonical mapping $\mathrm{Spec}(R) \to \mathrm{Max}(R)$ which associates to each prime ideal the unique maximal ideal containing it, $\overline{K}$ is clopen (see Exercise 9.4.14). Put $J_1 = \bigcap \{P \mid P \in \overline{K}\}$ and $J_2 = \bigcap \{Q \mid Q \in \mathrm{Spec}(R) \setminus \overline{K}\}$. As $J_1 \cap J_2 = \mathrm{nil}(R) = 0$, J_1 and J_2 are orthogonal ideals in R.

Since $\mathrm{Spec}(R)$ is compact and $\overline{K}$ is clopen, there exist $a_1, \ldots, a_n \in R$ and $b_1, \ldots, b_m \in R$ such that

$$\overline{K} = \bigcup_{i=1}^{n} U(a_i) \quad \text{and} \quad \mathrm{Spec}(R) \setminus \overline{K} = \bigcup_{j=1}^{m} U(b_j).$$

Equivalently,

$$\overline{K} = \bigcap_{j=1}^{m} V(b_j) \quad \text{and} \quad \text{Spec}(R) \setminus \overline{K} = \bigcap_{i=1}^{n} V(a_i).$$

As a result, $b_j \in P$ for every $P \in \overline{K}, 1 \le j \le m$ and $a_i \in Q$ for every $Q \in \text{Spec}(R) \setminus \overline{K}, 1 \le i \le n$. It follows that $\{a_1, \ldots, a_n, b_1, \ldots, b_m\} \subseteq J_1 + J_2$. Since

$$\text{Spec}(R) = \bigcup_{i=1}^{n} U(a_i) \cup \bigcup_{j=1}^{m} U(b_j)$$

there is no prime ideal of R which contains $J_1 + J_2$; in other words, $J_1 + J_2 = R$.

Take $e_1 \in J_1$ and $e_2 \in J_2$ such that $e_1 + e_2 = 1$. Then $e_1 = e_1(e_1 + e_2) = e_1^2$ so e_1 is an idempotent wherefore $e_2 = 1 - e_1$ is an idempotent as well. Take $x \in J_1$. Then $x = x(e_1 + e_2) = xe_1$ and hence $J_1 = Re_1$. Similarly $J_2 = Re_2 = R(1 - e_1)$. It follows that $\overline{K} = V(e_1)$ and thus, $K = V'(e_1)$, which was to show.

In the general case, that is, if $\text{nil}(R)$ is not necessarily zero, we apply the above argument to $R/\text{nil}(R)$. Since the spectra of R and $R/\text{nil}(R)$ are homeomorphic (e.g., [6, Exercise 21 in Chap. I] or [10, Remark on p. 101]), the assumptions apply, and since idempotents lift from $R/\text{nil}(R)$, see the result below, we obtain the claim. $\square$

Lemma 9.3.18 *Let R be a commutative unital ring. Then idempotents lift modulo the nil radical $\text{nil}(R)$.*

Proof Let $a \in R$ be such that $a - a^2 \in \text{nil}(R)$. Then, for some $n \in \mathbb{N}$, $(a - a^2)^n = 0$. The binomial formula yields

$$a^n - \binom{n}{1} a^{n-1} a^2 + \binom{n}{2} a^{n-2} a^4 - \cdots + a^{2n} = 0$$

whence $a^n - a^{n+1} b = 0$ for some $b \in R$. Put $e = (ab)^n$; then $a^n b^n = a^{n+1} b b^n = a^{n+1} b^{n+1} = \ldots = a^{n+n} b^{n+n} = a^{2n} b^{2n}$. Therefore, $e = e^2$ and

$$a + \text{nil}(R) = a^n + \text{nil}(R) = a^{n+1} b + \text{nil}(R) = (a^{n+1} + \text{nil}(R))(b + \text{nil}(R))$$

$$= (a + \text{nil}(R))(b + \text{nil}(R)) = ab + \text{nil}(R)$$

and hence

$$a + \text{nil}(R) = a^n + \text{nil}(R) = (a + \text{nil}(R))^n = (ab + \text{nil}(R))^n$$

$$= (ab)^n + \text{nil}(R) = e + \text{nil}(R)$$

so that the idempotent $a + \text{nil}(R)$ in $R/\text{nil}(R)$ lifts to the idempotent e in R. $\square$

9.4   Exercises

Exercise 9.4.1 Let M_1, M_2 and N be modules satisfying $M_1 \subseteq N \subseteq M_1 \oplus M_2$. Show that $N = M_1 \oplus (N \cap M_2)$.

Exercise 9.4.2 Use Theorem 9.1.4 to prove that, for modules M, N, L, K and $T \in R\text{-}\mathcal{M}od$, if $M = L \oplus K = N \oplus T$ and T has the finite exchange property and $N \leq L$, then K has the finite exchange property.

Exercise 9.4.3 Let R be an exchange ring and let $e \in R$ be an idempotent. Show that the subring eRe of R is an exchange ring. (The identity of eRe is e.)

Exercise 9.4.4 A unital ring R is said to be *semiregular* if $R/\text{rad}(R)$ is von Neumann regular and idempotents lift modulo $\text{rad}(R)$. Show that every semiregular ring is an exchange ring.

Exercise 9.4.5 Let R be a commutative unital ring. Show that, if R is an exchange ring, then R is clean.

Exercise 9.4.6 Let R be an exchange ring. Show by induction that, for every finite family $a_1, \ldots, a_n$ in R such that $\sum_{i=1}^{n} a_i = 1$, there exist orthogonal idempotents $e_i \in Ra_i$, $1 \leq i \leq n$ such that $\sum_{i=1}^{n} e_i = 1$.

Exercise 9.4.7 Use Theorems 9.1.4 and 9.2.6 to show the following. Let e be an idempotent in a unital ring R. Then R is an exchange ring if and only if eRe and $(1 - e)R(1 - e)$ are exchange rings.

Exercise 9.4.8 Show directly from the definition that every commutative exchange ring is a Gelfand ring.

Exercise 9.4.9 For a commutative unital ring R, let $\text{E}(R) = \{U'(e) \mid e = e^2 \in R\}$; compare p. 123. Show that $\text{E}(R)$ is closed under finite intersections and unions as well as complements.

Exercise 9.4.10 Let R be a commutative unital ring. Show directly from the definitions that, if R is a tb-ring, then R is an exchange ring.

Exercise 9.4.11 Let R be a commutative unital ring. Show directly from the definitions that, if R is a clean ring, then R is a tb-ring.

Exercise 9.4.12 Use Proposition 5.2.4 to show that every Artinian commutative ring is a pm-ring.

Exercise 9.4.13 A commutative unital ring R is called *local* if it contains a unique maximal ideal. Show that every such ring is a *pm*-ring.

Exercise 9.4.14 Suppose R is a *pm*-ring. Show that the canonical mapping $\text{Spec}(R) \rightarrow \text{Max}(R)$ which associates to each prime ideal the unique maximal ideal containing it is continuous.

Semiprimitivity of Group Rings

10

The starting point for this chapter is the discussion around Maschke's theorem, Theorem 7.2.1, which provides us with conditions under which the group ring $K[G]$ is semisimple, provided G is a finite group. For any field K, the elements of G form a basis of the K-vector space $K[G]$ and if the ring $K[G]$ is semisimple, then it is necessarily Artinian, hence finite dimensional (Corollary 6.2.5 and Exercise 5.5.6). As a result, we cannot expect $K[G]$ to be semisimple for an infinite group G.

The 'next best' property one can expect for infinite groups therefore is semiprimitivity, see also Proposition 7.3.7. An important and still open problem in ring theory, which was popularised by Kaplansky, asks for characterisations of those groups for which $K[G]$ is semiprimitive. (This was known as the "semisimplicity problem" before the terminology changed.) The larger part of this chapter will be devoted to a study of some of the known results. In our approach, we will use techniques from functional analysis (developed in detail in Sect. 10.2) to obtain a positive answer when $K = \mathbb{C}$, the field of complex numbers. This is a nice illustration of how different areas of Mathematics can work together.

Throughout this chapter, K will denote a field and groups will be written multiplicatively.

10.1 Basic Properties

We begin by establishing some basic properties of group rings. For the definition of the operations, see (ix) in Examples 1.2.

Proposition 10.1.1 *Let G be a finitely generated torsion-free abelian group. Then, for some $n \in \mathbb{N}$, $K[G]$ is contained isomorphically between the polynomial ring $K[x_1, \ldots, x_n]$ and the rational field $K(x_1, \ldots, x_n)$. In particular, $K[G]$ is an integral domain.*

Proof The fundamental theorem of abelian groups yields a decomposition $G = \langle g_1 \rangle \times \ldots \times \langle g_n \rangle$ as a product of infinite cyclic groups, for some natural number n. The canonical homomorphism $K[x_1, \ldots, x_n] \longrightarrow K[G]$ given by $x_i \mapsto g_i$, $1 \leq i \leq n$ is an embedding since every element of G is uniquely of the form $g_1^{m_1} \cdots g_n^{m_n}$. Under this embedding, for each $\alpha \in K[G]$, there is $m \in \mathbb{N}$ such that $(x_1, \ldots, x_n)^m \alpha \in K[x_1, \ldots, x_n]$ and hence, $K[G]$ is an integral domain. As such, it is canonically contained in the field of fractions $K(x_1, \ldots, x_n)$. $\qquad \square$

As a consequence we obtain the following result.

Corollary 10.1.2 *Let G be an infinite cyclic group. Then $K[G]$ is a principal ideal domain.*

Proof By the above proposition, we already know that $K[G]$ is an integral domain. Let I be a non-zero ideal of $K[G]$. Let $g \in G$ be a generator. Then every element of $K[G]$ is of the form $g^k f(g)$ for some $k \in \mathbb{Z}$ and a polynomial $f(g) \in K[G]$. Suppose $g^k f(g) \in I$; then $f(g) = g^{-k} g^k f(g) \in I$. Take a non-zero polynomial $t(g) \in I$ of minimal degree and divide $f(g)$ by $t(g)$. By the division algorithm and the minimality assumption we conclude that $t(g)$ divides $f(g)$ and thus $t(g)$ generates the ideal I. Consequently, $K[G]$ is a principal ideal domain. $\qquad \square$

Suppose H is a subgroup of the group G; then, clearly, $K[H]$ is a unital subring of $K[G]$. We have the following relation between the Jacobson radicals of these two rings.

Proposition 10.1.3 *Let H be a subgroup of G. Then*

$$\mathrm{rad}(K[G]) \cap K[H] \subseteq \mathrm{rad}(K[H]).$$

We prepare the proof of this proposition by the following lemma.

Lemma 10.1.4 *Let R be a unital subring of the unital ring S. Suppose, as left R-modules, R is a direct summand of S. Then $\mathrm{rad}(S) \cap R \subseteq \mathrm{rad}(R)$.*

Proof Suppose that $_R S = {_R R} \oplus {_R M}$ for some left R-module $_R M$. Take $\alpha \in \mathrm{rad}(S) \cap R$ and fix $\beta \in R$. Since there is $s \in S$ such that $(1 - \alpha\beta)s = 1 = s(1 - \alpha\beta)$ and s can be written (uniquely) as $s = r + m$ with $r \in R$, $m \in M$, we have

$$1 = (1 - \alpha\beta)r + (1 - \alpha\beta)m.$$

As α belongs to R and R is a unital subring, this entails that $1 = (1 - \alpha\beta)r$ and $(1 - \alpha\beta)m = 0$. Upon multiplying

$$1 = r(1 - \alpha\beta) + m(1 - \alpha\beta)$$

on the right by r we obtain

$$r = r(1 - \alpha\beta)r + m(1 - \alpha\beta)r = r + m$$

and hence $m = 0$. As a result, $1 - \alpha\beta$ is invertible in R and so $\alpha \in \text{rad}(R)$, by Proposition 7.3.2. □

Proof of Proposition 10.1.3 The set $M = \{\alpha \in K[G] \mid a_g = 0 \text{ for all } g \in H\}$ is a left $K[H]$-submodule of $K[G]$ and $K[G] = K[H] \oplus {}_RM$. By Lemma 10.1.4, the assertion follows. □

We turn our attention to fields with finite characteristic. Suppose K is a field with characteristic $p > 0$ and let A be an algebra over K. If A is commutative, the power map $\alpha \mapsto \alpha^p$ is a homomorphism, as is easily checked. In the general case, we have the following generalisation. We shall denote by $[A, A]$ the subspace of A spanned by all commutators $[\alpha, \beta] = \alpha\beta - \beta\alpha$.

Lemma 10.1.5 *Let K be a field with characteristic $p > 0$, and for given $n \in \mathbb{N}$, set $q = p^n$. Let A be an algebra over K and let $\alpha_1, \ldots, \alpha_m \in A$. Then, for some $\beta \in [A, A]$, we have*

$$(\alpha_1 + \cdots + \alpha_m)^q = \alpha_1^q + \cdots + \alpha_m^q + \beta. \tag{10.1.1}$$

Proof We have $(\alpha_1 + \cdots + \alpha_m)^q = \alpha_1^q + \cdots + \alpha_m^q + \beta$, where β is the sum of all words $\alpha_{i_1} \cdots \alpha_{i_q}$ with at least two distinct indices. Let C_q be the cyclic group of order q. Suppose $\omega_1 = \alpha_{i_1} \cdots \alpha_{i_q}$ and $\omega_2 = \alpha_{i_j} \cdots \alpha_{i_q}\alpha_{i_1} \cdots \alpha_{i_{j-1}}$ are cyclic permutations of each other. Then,

$$\omega_1 - \omega_2 = (\alpha_{i_1} \cdots \alpha_{i_{j-1}})(\alpha_{i_j} \cdots \alpha_{i_q}) - (\alpha_{i_j} \cdots \alpha_{i_q})(\alpha_{i_1} \cdots \alpha_{i_{j-1}}) \in [A, A].$$

Hence, modulo $[A, A]$, all cyclic permutations of a word are equal and the number of formally distinct permutations of a word which occur in β as above is the size of a non-trivial orbit in C_q, therefore divisible by p. This proves the assertion. □

This result will become handy in Sect. 10.3 below.

10.2 Some Analytic Structure on $\mathbb{C}[G]$

Let G be an arbitrary group. When G is infinite, $\mathbb{C}[G]$ is an infinite-dimensional vector space over the complex numbers and it is expedient to introduce some analytic structure on it. Firstly, we see that it is an inner product space in a canonical way.

Definition 10.2.1 Let $\alpha = \sum_{g \in G} a_g g$, $\beta = \sum_{g \in G} b_g g$ belong to $\mathbb{C}[G]$. We put

$$(\alpha \mid \beta) = \sum_{g \in G} a_g \bar{b}_g$$

(where $\bar{b}$ denotes the complex conjugate of $b \in \mathbb{C}$) and call this the *canonical inner product* on $\mathbb{C}[G]$.

It will be shown in Exercise 10.4.1 that the above indeed defines an inner product on $\mathbb{C}[G]$ and that $\|\alpha\|_2 = (\alpha \mid \alpha)^{1/2}$ therefore is a norm on $\mathbb{C}[G]$. However, the associated metric space is in general not complete. In other words, $(\mathbb{C}[G], (\cdot \mid \cdot))$ is in general not a Hilbert space.

Beside the structure of an inner product space, $\mathbb{C}[G]$ also has the structure of a *normed algebra*.

Definition 10.2.2 Let $\alpha = \sum_{g \in G} a_g g \in \mathbb{C}[G]$. We put $\|\alpha\|_1 = \sum_{g \in G} |a_g|$, where $|a|$ of course denotes the absolute value of the complex number a.

Proposition 10.2.3 *The above definition of $\|\alpha\|_1$ turns $\mathbb{C}[G]$ into a normed space. Moreover, for all α, $\beta \in \mathbb{C}[G]$, we have $\|\alpha\beta\|_1 \le \|\alpha\|_1 \|\beta\|_1$ and $\mathbb{C}[G]$ is a complex normed algebra with identity 1.*

Proof It is straightforward to check that, for all α, $\beta \in \mathbb{C}[G]$, $\|\alpha + \beta\|_1 \le \|\alpha\|_1 + \|\beta\|_1$, that $\|c\alpha\|_1 = |c| \|\alpha\|_1$ for every $c \in \mathbb{C}$ and that $\|\alpha\|_1 = 0$ implies that $\alpha = 0$. Thus, $\| \cdot \|_1$ defines a norm on $\mathbb{C}[G]$.

From the triangle inequality and the fact that $\|\alpha g\|_1 = \|\alpha\|_1$ we obtain

$$\|\alpha\beta\|_1 = \left\| \sum_{g \in G} \alpha b_g g \right\|_1 \le \sum_{g \in G} \|\alpha b_g g\|_1 = \sum_{g \in G} \|\alpha\|_1 |b_g| = \|\alpha\|_1 \|\beta\|_1.$$

As a result, $(\mathbb{C}[G], \| \cdot \|_1)$ is a normed algebra. Clearly, 1 is a multiplicative identity.
$\square$

As all good things come in threes, we will introduce another norm on $\mathbb{C}[G]$. Let $H = \ell^2(G)$ be the Hilbert space obtained from $(\mathbb{C}[G], (\cdot \mid \cdot))$, compare Exercise 10.4.1. The algebra $\mathbb{C}[G]$ acts on H in a canonical way: for clarity let us denote the elements in H by ξ so that $\xi \in \mathbb{C}[G]$ can be written as $\xi = \sum_{g \in G} x_g g$ with $x_g \in \mathbb{C}$. For such ξ and $\alpha \in \mathbb{C}[G]$ we have $\alpha\xi \in \mathbb{C}[G] \subseteq H$, that is, $\mathbb{C}[G]$ acts as left multiplication operator. The above inequalities imply that this action is continuous, so can be extended to all of H. Moreover, since $\|\alpha\xi\|_2 \le \|\alpha\|_2 \|\xi\|_2$ for all $\xi \in H$, the action is injective ($\alpha 1 = 0 \Rightarrow \alpha = 0$) and it makes sense to define our third norm as

$$\|\alpha\|_{op} = \sup\{\|\alpha\xi\|_2 \mid \xi \in H, \ \|\xi\|_2 = 1\}.$$

Because $\mathbb{C}$ has a distinguished conjugation $z \mapsto \bar{z}$, we can define an *involution*, that is, an anti-linear automorphism $\alpha \mapsto \alpha^*$ on $\mathbb{C}[G]$ in the following way:

$$\text{for } \alpha = \sum_{g \in G} a_g g \text{ we put } \alpha^* = \sum_{g \in G} \bar{a}_g g^{-1}.$$

Proposition 10.2.4 *For all $\alpha, \beta, \gamma \in \mathbb{C}[G]$, we have $(\alpha + \beta)^* = \alpha^* + \beta^*$, $(\alpha\beta)^* = \beta^*\alpha^*$ and $(\alpha^*)^* = \alpha$. Furthermore,*

$$(\alpha\gamma^* \mid \beta) = (\alpha \mid \beta\gamma) = (\beta^*\alpha \mid \gamma). \tag{10.2}$$

Proof The first three identities are very easy to check. Next we recall the definition of the trace on $\mathbb{C}[G]$ from Exercise 1.3.9: $\text{tr} \left(\sum_{g \in G} a_g g \right) = a_1$ and, therefore, $(\alpha \mid \beta) = \text{tr}\,\alpha\beta^*$. As a consequence, for all $\alpha, \beta, \gamma \in \mathbb{C}[G]$,

$$(\alpha\gamma^* \mid \beta) = \text{tr}\,\alpha\gamma^*\beta^* = \text{tr}\,\alpha(\beta\gamma)^* = (\alpha \mid \beta\gamma)$$
$$= \text{tr}\,(\beta\gamma)^*\alpha = \text{tr}\,\gamma^*\beta^*\alpha = (\beta^*\alpha \mid \gamma),$$

where of course we used the trace property. □

In the language of bounded linear operators on the Hilbert space H, α^* is the adjoint of α: $(\xi \mid \alpha\eta) = (\alpha^*\xi \mid \eta)$, by (10.2), compare Exercise 10.4.4.

The "operator norm" has the additional nice property that $\|\alpha\|_{op} = \|\alpha^*\|_{op}$ and that $\|\alpha^*\alpha\|_{op} = \|\alpha\|_{op}^2$. With the involution $\alpha \mapsto \alpha^*$, the completion of $(\mathbb{C}[G], \| \cdot \|_{op})$ becomes what is known as a *C*-algebra*. See also Example 3.2.4 and Exercises 10.4.4 and 10.4.5.

The C*-algebra which is obtained in this way is the *reduced group C*-algebra* of G, denoted by $C_r^*(G)$. A special property of this kind of C*-algebra is that it carries a faithful trace. The trace tr on $\mathbb{C}[G]$ satisfies

$$|\text{tr}\,\alpha|^2 \leq \|\alpha\|_2^2 = (\alpha 1 \mid \alpha 1) = (\alpha^*\alpha 1 \mid 1) \leq \|\alpha^*\alpha\|_{op} = \|\alpha\|_{op}^2$$

so that tr is a bounded linear functional and thus can be extended from the dense subspace $\mathbb{C}[G]$ to the whole algebra $C_r^*(G)$. We shall denote this extended functional by τ.

Lemma 10.2.5 *The functional τ is a normalised faithful trace on $C_r^*(G)$, that is, $\tau(1) = 1$, $\tau(ab) = \tau(ba)$ for all $a, b \in C_r^*(G)$ and $\tau(a^*a) \geq 0$ with $\tau(a^*a) = 0$ if and only if $a = 0$.*

Proof By definition, $\tau(1) = \text{tr}\,1 = 1$ and since tr is a bounded linear functional on the dense subspace $\mathbb{C}[G]$ and the multiplication in $C_r^*(G)$ is continuous, it follows that $\tau(ab) = \tau(ba)$ for all $a, b \in C_r^*(G)$. Let $a \in C_r^*(G)$. Let $\varepsilon > 0$ and take

$\alpha \in \mathbb{C}[G]$ such that $\|a - \alpha\|_{op} < \varepsilon$. Then

$$\|a^*a - \alpha^*\alpha\|_{op} \leq \|a^* - \alpha^*\|_{op}\|\alpha\|_{op} + \|a^*\|_{op}\|a - \alpha\|_{op}$$
$$< \varepsilon(\varepsilon + \|a\|_{op}) + \|a\|_{op}\varepsilon = \varepsilon(\varepsilon + 2\|a\|_{op}) = \varepsilon'.$$

Since $\operatorname{tr}\alpha^*\alpha = \sum_{g \in G} |a_g|^2 \geq 0$, for $\varepsilon' > 0$ such that $\operatorname{tr}\alpha^*\alpha - \varepsilon' \geq 0$, we obtain

$$\tau(a^*a) = \operatorname{tr}\alpha^*\alpha - \tau(\alpha^*\alpha - a^*a) \geq \operatorname{tr}\alpha^*\alpha - \|a^*a - \alpha^*\alpha\|_{op} > \operatorname{tr}\alpha^*\alpha - \varepsilon' \geq 0.$$

In the case where $\tau(a^*a) = 0$ the above inequality yields $\operatorname{tr}\alpha^*\alpha = 0$, that is, $\alpha = 0$, and as a result, a must be zero too. Therefore the trace τ is faithful.   □

The next lemma is a useful general result in C^*-algebras and can be found in many of the standard textbooks. We provide a proof for the convenience of the reader.

Lemma 10.2.6 *Let A be a unital C^*-algebra. Every idempotent in A is equivalent to a projection in A. That is, if $e \in A$ satisfies $e = e^2$ then there is $p \in A$ with $p = p^2 = p^*$ such that, for some $x, y \in A$, we have $e = xy$ and $p = yx$.*

Proof We first observe that every element of the form $1 + b^*b$, $b \in A$ is invertible in A. One of the many ways to see this is to consider the unital C^*-subalgebra of A generated by 1 and b^*b. Since b^*b is selfadjoint, this C^*-subalgebra is commutative and therefore isomorphic to $C(X)$ for a compact Hausdorff space X; compare Examples 3.2.2 and 3.2.4. As a function on X, $1 + b^*b$ is of the form $1 + \bar{f}f$ and this function is certainly invertible as it has no zeros. Consequently, the (isomorphic image of the) inverse serves as the inverse of $1 + b^*b$ in A.

We now set $a = 1 + (e - e^*)^*(e - e^*)$ for our given idempotent e. We compute

$$ea = e + e(e^*e - e - e^* + ee^*) = ee^*e = e + (e^*e - e - e^* + ee^*)e = ae,$$

that is, e and a commute and hence e and a^{-1} commute. We put $p = ee^*a^{-1}$ and observe that $p^* = a^{-1}ee^* = p$ (since a is selfadjoint, so is its inverse). Furthermore,

$$p^2 = p^*p = a^{-1}ee^*ee^*a^{-1} = a^{-1}aee^*a^{-1} = p$$

so that p is a projection. Finally, we have $ep = p$ and $pe = p^*e = a^{-1}ee^*e = a^{-1}ae = e$ so that e and p are indeed equivalent.   □

The above observations enable us to draw a consequence on the range of the trace on idempotents in $C_r^*(G)$. Let $e \in C_r^*(G)$ be an idempotent and let p be a projection equivalent to e. Then $\tau(e) = \tau(p) = \tau(p^*p) \geq 0$ and equal to 0 only if $e = 0$ (Lemma 10.2.5). Applying the same reasoning to $1 - e$ we find that $\tau(e) = 1$

if and only if $e = 1$. This can be used to determine an algebraic property of the group ring $\mathbb{C}[G]$.

Recall from Exercise 4.3.2 that a unital ring R is *von Neumann finite*, if, for $x, y \in R$, $xy = 1$ implies that $yx = 1$.

Theorem 10.2.7 *For every group G, the group ring $\mathbb{C}[G]$ is von Neumann finite.*

Proof Let $x, y \in \mathbb{C}[G]$ satisfy $xy = 1$; then $e = yx$ is an idempotent different from 0. Since $\operatorname{tr} e = \operatorname{tr} xy = 1$ if follows that $e = 1$, i.e., $yx = 1$. $\square$

This theorem in fact holds for all groups rings $K[G]$ where K has characteristic zero. A result by Kaplansky from 1969, it can be derived from the above special case, or by other means, see, e.g., [31, Corollary 2.1.9].

10.3 The Semiprimitivity Problem

A primitive ring is one that has a faithful simple module; according to the convention in this book, we focus on the left-handed version (compare Definition 7.3.4). As primitive rings are somewhat scarce, one wants to study more general rings by looking at *all* their simple modules (equivalently, irreducible representations) but for this to work, the canonical obstruction, the Jacobson radical, has to vanish. These are the *semiprimitve* rings (Definition 7.3.4).

When is the group ring $K[G]$ semiprimitive? The answer to this question will depend on properties of the field K and the group G and possibly on the interaction of the two. We will discuss this question for infinite groups in this section.

The first case we investigate is the group ring with complex numbers as coefficients, this will lead us to one of the first results on the above question, originally obtained by Rickart in 1950. We present a modified version of his original proof which uses functional analytic techniques which we already prepared ourselves for in the previous section.

Theorem 10.3.1 *For every group G, the group ring $\mathbb{C}[G]$ is semiprimitive.*

Proof When G is finite, $\mathbb{C}[G]$ is semisimple by Maschke's theorem (Theorem 7.2.1), hence semiprimitive. For the case when G is infinite, we will use the Banach algebra $\ell^1(G)$ discussed in Sect. 10.2 and Exercise 10.4.2. The strategy of the proof is as follows. For $\alpha \in \operatorname{rad}(\mathbb{C}[G])$, we will construct an entire function F such that $F(1) = \sum_{n=0}^{\infty} \operatorname{tr} \alpha^n$; this will entail that $\lim_{n \to \infty} \operatorname{tr} \alpha^n = 0$. On the other hand, we will show that the only $\alpha \in \operatorname{rad}(\mathbb{C}[G])$ that allows for the construction of the above-mentioned F is $\alpha = 0$, which will complete the argument that $\mathbb{C}[G]$ is semiprimitive.

Let us prove the last assertion first. Take $\beta \in \operatorname{rad}(\mathbb{C}[G]) \setminus \{0\}$ and put $\alpha = \frac{\beta \beta^*}{\|\beta\|_2^2}$, where we use the notation of Sect. 10.2. Then $\alpha^* = \alpha \in \operatorname{rad}(\mathbb{C}[G])$ since the

Jacobson radical is an ideal of $\mathbb{C}[G]$. It follows that

$$\operatorname{tr} \alpha = \frac{\operatorname{tr} \beta \beta^*}{\|\beta\|_2^2} = \frac{\|\beta\|_2^2}{\|\beta\|_2^2} = 1.$$

Suppose $\operatorname{tr} \alpha^{2^m} \geq 1$ for some $m \in \mathbb{N}_0$. Then

$$\operatorname{tr} \alpha^{2^{m+1}} = \operatorname{tr} \alpha^{2^m} (\alpha^{2^m})^* = \|\alpha^{2^m}\|_2^2 \geq |\operatorname{tr} \alpha^{2^m}|^2 \geq 1;$$

thus, by induction, $\operatorname{tr} \alpha^{2^m} \geq 1$ for all $m \in \mathbb{N}_0$ and therefore the sequence $(\operatorname{tr} \alpha^n)_{n \in \mathbb{N}}$ alluded to above cannot tend to zero.

We shall now go about the construction of the function F. Since, for every $z \in \mathbb{C}$, $1 - z\alpha$ is invertible in $\mathbb{C}[G]$ (Proposition 7.3.2), we can define $\phi \colon \mathbb{C} \to \mathbb{C}[G]$ by $\phi(z) = (1 - z\alpha)^{-1}$. Suppose $\alpha \neq 0$ and that $|z| < \frac{1}{\|\alpha\|_1}$. In this case the series $\sum_{n=0}^{\infty} (z\alpha)^n$ converges with limit $(1 - z\alpha)^{-1}$ (the proof is the same as for the usual geometric series in Real Analysis, see Exercise 10.4.3).

We define $F \colon \mathbb{C} \to \mathbb{C}$ by $F(z) = \operatorname{tr} \phi(z)$. On the open disk $\{z \in \mathbb{C} \mid |z| < \frac{1}{\|\alpha\|_1}\}$, F is given by the power series $F(z) = \sum_{n=0}^{\infty} z^n \operatorname{tr} \alpha^n$, since the trace is linear and continuous. We show that F is entire; then the identity theorem entails that F is given by this power series on the entire complex plane, in particular, $F(1) = \sum_{n=0}^{\infty} \operatorname{tr} \alpha^n$ which was to prove.

In order to show that F is differentiable at every $z \in \mathbb{C}$, note first that $\phi(z)$ and $\phi(w)$ commute for all $z, w \in \mathbb{C}$. Therefore,

$$\phi(z) - \phi(w) = (1 - z\alpha)^{-1} - (1 - w\alpha)^{-1}$$

$$= ((1 - w\alpha) - (1 - z\alpha))(1 - z\alpha)^{-1}(1 - w\alpha)^{-1}$$

$$= (z - w)\alpha\, \phi(z)\phi(w).$$

Taking traces yields, for $z \neq w$,

$$\frac{F(z) - F(w)}{z - w} = \operatorname{tr}(\alpha\, \phi(z)\phi(w)),$$

thus, by continuity of ϕ and the trace,

$$\lim_{w \to z} \frac{F(z) - F(w)}{z - w} = \operatorname{tr}(\alpha\, \phi(z)^2).$$

As a result, the derivative of F at z exists and is equal to $\operatorname{tr}(\alpha\, \phi(z)^2)$. □

There do exist purely algebraic proofs of Theorem 10.3.1 but it is instructive to see how Analysis can be used to obtain an algebraic result.

From Lemma 7.3.8 we know that $\mathrm{rad}(K[G])$ contains every nil ideal of $K[G]$. This is a starting point to obtain another case where $K[G]$ is semiprimitive.

Lemma 10.3.2 *In every group ring, each algebraic element in the radical is nilpotent.*

Proof Let $\alpha \in \mathrm{rad}(K[G])$ and suppose it satisfies a polynomial equation over K which we write as

$$\alpha^n(1 + c_1\alpha + c_2\alpha^2 + \cdots + c_r\alpha^r) = 0$$

for suitable $r, n \in \mathbb{N}_0$ and $c_i \in K$. Set $\gamma = c_1\alpha + c_2\alpha^2 + \cdots + c_r\alpha^r \in \mathrm{rad}(K[G])$; then $1 + \gamma$ is invertible and thus $\alpha^n = 0$. $\square$

Lemma 10.3.3 *Let G be a group and let K be a field with the property that $|K| > \dim_K K[G]$ (as a strict inequality of possibly infinite cardinal numbers). Then every element in $\mathrm{rad}(K[G])$ is algebraic.*

Proof Let $\alpha \in \mathrm{rad}(K[G])$; since $1 - c\alpha$ is invertible for every $c \in K$ (Proposition 7.3.2), and $\dim_K K[G]$ is strictly less than the cardinality of K, there have to be finitely many $a_i, c_i \in K \setminus \{0\}$, $1 \leq i \leq n$, for some $n \in \mathbb{N}$, such that

$$a_1(1 - c_1 1\alpha)^{-1} + a_2(1 - c_2\alpha)^{-1} + \cdots + a_n(1 - c_n\alpha)^{-1} = 0.$$

Upon multiplying the above identity by $(1 - c_1\alpha) \cdots (1 - c_n\alpha)$ we obtain a non-zero polynomial over K of which α is a root. $\square$

From the two results above we see that, for very large coefficient fields, every element in the radical is nilpotent; hence the radical itself is a nil ideal.

Theorem 10.3.4 *Let K be an uncountable field and let G be a group. Then $\mathrm{rad}(K[G])$ is a nil ideal.*

Proof Let $\alpha = \sum_{g \in G} a_g g \in \mathrm{rad}(K[G])$ and let H be a finitely generated subgroup of G which contains $\{g \in G \mid a_g \neq 0\}$. Then $\dim_K K[H]$ is countable so strictly smaller than $|K|$. By Lemma 10.3.3 together with Lemma 10.3.2, $\mathrm{rad}(K[H])$ is a nil ideal. As $\mathrm{rad}(K[G]) \cap K[H] \subseteq K[H]$ (Proposition 10.1.3), it follows that $\alpha \in \mathrm{rad}(K[G]) \cap K[H]$ is nilpotent, so altogether $K[G]$ is a nil ideal. $\square$

Thus the remaining task is to determine when $K[G]$ has no non-zero nil ideals.

Theorem 10.3.5 *Let K be a field with characteristic $p > 0$ and let G be a group. For $\alpha = \sum_{g \in G} a_g g \in K[G]$, let*

$$p\text{-supp}\,\alpha = \{g \in G \mid a_g \neq 0 \text{ and } g^{p^i} = 1 \text{ for some } i \in \mathbb{N}\}$$

be the p-support of α. Suppose α is nilpotent. Then

$$\operatorname{tr}\alpha + \Big(\sum_{g \in p\text{-supp}\,\alpha} a_g \Big) = 0.$$

Therefore, either $\operatorname{tr}\alpha = 0$ or $p\text{-supp}\,\alpha \neq \emptyset$.

Proof Since α is nilpotent, there is $n_0 \in \mathbb{N}$ such that $\alpha^{p^n} = 0$ for all $n \geq n_0$. Moreover, as $p\text{-supp}\,\alpha$ is finite, we can choose n such that $g^{p^n} = 1$ for all $g \in p\text{-supp}\,\alpha$. Put $q = p^n$. Applying Lemma 10.1.5 we obtain

$$0 = \alpha^q = \sum_{g \in G} a_g^q\, g^q + \beta$$

for some $\beta \in \big[K[G], K[G] \big]$. Since the trace vanishes on commutators this entails that

$$0 = \operatorname{tr}\alpha^q = \sum_{g^q = 1} a_g^q = \Big(\sum_{g^q = 1} a_g \Big)^q.$$

By the choice of n, $g^q = 1$ if and only if $g \in p\text{-supp}\,\alpha$ or $g = 1$. Thus, $\operatorname{tr}\alpha + \Big(\sum_{g \in p\text{-supp}\,\alpha} a_g \Big) = 0$ as claimed. □

Let p be a prime. A group G is called a p'-group if, for all $g \in G$, $g \neq 1$ and all $i \in \mathbb{N}$, $g^{p^i} \neq 1$. We now combine the last two theorems to obtain an answer to the semiprimitivity problem for certain fields and certain groups.

Theorem 10.3.6 *Let K be an uncountable field with characteristic $p > 0$ and let G be a p'-group. Then $K[G]$ is semiprimitive.*

Proof By Theorem 10.3.4, it suffices to show that the only nilpotent element in $\operatorname{rad}(K[G])$ is the zero element. Let $\alpha = \sum_{g \in G} a_g g \in \operatorname{rad}(K[G])$ be nilpotent. Then $a_g = \operatorname{tr}(\alpha g^{-1}) = 0$, by Theorem 10.3.5, since G is a p'-group. This proves the claim. □

Remark 10.3.7 The above result, due to Amitsur (1957), also holds for fields with characteristic 0 and arbitrary groups. To this end, one has to extend the previous two theorems to this setting and add quite a bit of additional terminology and techniques. Since these would take us here too far afield, we refer to [31, Chap. 7] for the details.

A wealth of information on the semiprimitivity problem is contained in Passman's book [31] as well as in [24]. On the other hand, the following fundamental question remains open:

Is the group ring $\mathbb{Q}[G]$ semiprimitive for every group G?

Remark 10.3.8 The three famous 'Kaplansky conjectures' on the structure of group rings are as follows. Let K be a field and G be a torsion-free group. Then

(U) $K[G]$ has no non-trivial units, that is, every element in $K[G]^\times$ is of the form ag with $a \in K^*$ and $g \in G$.
(Z) $K[G]$ has no non-trivial zero divisors, that is, every zero divisor is equal to 0.
(I) $K[G]$ has no non-trivial idempotents, that is, every idempotent in $K[G]$ is either equal to 1 or to 0.

The relation between them is

$$(U) \Longrightarrow (Z) \Longrightarrow (I).$$

The second implication is easy: every idempotent $e \in K[G] \setminus \{0, 1\}$ yields non-trivial zero divisors via $e(1 - e) = e - e^2 = 0$. The first is a bit harder and uses Connell's result ([31, Theorem 4.2.10]) that $K[G]$ is a prime ring under the given assumptions. They were open for many decades and proven in a large number of special cases. Only in 2021, conjecture (U) was settled in the negative by Gardam [18].

Since these questions are related to other important problems, for example, the Baum–Connes conjecture in operator algebras, there is still quite some activity in the study of group rings.

10.4 Exercises

Exercise 10.4.1 Show that the mapping $(\alpha, \beta) \mapsto (\alpha \mid \beta)$ as defined in Definition 10.2.1 gives $\mathbb{C}[G]$ the structure of an inner product space and therefore $\|\alpha\|_2 = (\alpha \mid \alpha)^{1/2}$ yields a norm on $\mathbb{C}[G]$. Show further that the associated metric $(\alpha, \beta) \mapsto \|\alpha - \beta\|_2$ turns $\mathbb{C}[G]$ into a metric space which in general (i.e., for infinite groups) is not complete. Its completion to a Hilbert space is usually denoted by $\ell^2(G)$. (For the terminology, see any basic book on Functional Analysis such as [1, 14] or [26].)

Exercise 10.4.2 Prove that the normed algebra $\mathbb{C}[G]$ as described in Proposition 10.2.3 is in general not complete. Its completion to a Banach algebra is generally denoted by $\ell^1(G)$ and called an "L^1-group algebra". (You find the definition of $\ell^1(G)$ in any of the above-cited books.)

Exercise 10.4.3 Let A be a unital Banach algebra. Show that, for every $b \in A$ with $\|b\| < 1$, the series $\sum_{n=0}^{\infty} b^n$ converges and has limit $(1-b)^{-1}$. (This series is called the *Neumann series* after Carl Neumann. See, e.g., [1, Sect. 4.4].)

Exercise 10.4.4 Let H be a complex Hilbert space. A linear mapping $T \colon H \to H$ is called *bounded* if, for some constant $\mu \geq 0$ and all $\xi \in H$, the inequality $\|T\xi\|_2 \leq \mu \|\xi\|_2$ holds (we continue to denote the norm on H by $\|\cdot\|_2$). In this case, one can define the *operator norm* of T as

$$\|T\|_{op} = \sup\{\|T\xi\|_2 \mid \xi \in H, \|\xi\|_2 = 1\}.$$

Show that this indeed defines a norm on the complex vector space $B(H)$ of all bounded linear mappings on H. Show also that, if $S, T \in B(H)$, then $\|ST\|_{op} \leq \|S\|_{op}\|T\|_{op}$ and that $\|T\|_{op} = \|T^*\|_{op}$ and $\|T^*T\|_{op} = \|T\|_{op}^2$, where the *adjoint* T^* of T is given by the formula $(T\xi \mid \eta) = (\xi \mid T^*\eta)$ for all $\xi, \eta \in H$. Also prove that $B(H)$ with this norm is complete. (If you need help with this exercise, look into [1, Sect. 2.14], [26, Chap. 3, Sect. 9] or [32, Sect. 3.2].)

Exercise 10.4.5 A *C*-algebra* is a subalgebra of some $B(H)$ (as in the previous exercise) which contains with every operator also its adjoint and which is closed under convergence with respect to the operator norm. Suppose $\{A_i \mid i \in I\}$ is a family of C*-algebras contained in (a fixed) $B(H)$. Show that the intersection $\bigcap A_i$ is a C*-algebra in $B(H)$. Consequently, for every non-empty subset S of $B(H)$ there is a smallest C*-algebra in $B(H)$ containing S. (It is called the *C*-algebra generated by S*.)

Exercise 10.4.6 Let K be a subfield of $\mathbb{C}$ which is closed under complex conjugation and let G be a group. Use the involution $*$ on $K[G]$ (compare Proposition 10.2.4) to show that $K[G]$ does not contain any non-zero nil ideal.

Exercise 10.4.7 For a group G, let $\mathscr{F}(G)$ denote the set of all its finitely generated subgroups. Let $\alpha \in K[G]$, for some field K. Show that $\alpha \in \mathrm{rad}(K[G])$ if and only if $\alpha \in \mathrm{rad}(K[H])$ for all $H \in \mathscr{F}(G)$.

Exercise 10.4.8 Use Proposition 10.1.3 together with the previous exercise to show that, for a group G, $K[G]$ is semiprimitive if $K[H]$ is semiprimitive for all $H \in \mathscr{F}(G)$, and that $\mathrm{rad}(K[G])$ is a nil ideal if $\mathrm{rad}(K[H])$ is nil for all $H \in \mathscr{F}(G)$.

Exercise 10.4.9 Let R be a non-zero ring and let G be an infinite group. Define the *augmentation map* $\epsilon \colon R[G] \to R$ by $\epsilon(r) = r$ for all $r \in R$ and $\epsilon(g) = 1$ for all

$g \in G$. Assuming that $R[G]$ is semisimple, ker ϵ will be a direct summand which can be written as Re for some idempotent $e \in R$. Show that the idempotent $1 - e$ has to involve every group element which is impossible as G is infinite. Conclude that $R[G]$ can never be semisimple.

Bibliography

1. Allan, G.R.: Introduction to Banach Spaces and Algebras. Oxford Graduate Texts in Mathematics, vol. 20. Oxford University Press, Oxford (2011)
2. Anderson, F.W., Fuller, K.R.: Rings and Categories of Modules. Graduate Texts in Mathematics, vol. 13. Springer-Verlag, New York (1992)
3. Ara, P.: Extensions of exchange rings. J. Algebra **197**, 409–423 (1997)
4. Ara, P., Goodearl, K.R., O'Meara, K.C., Pardo, E.: Separative cancellation for projective modules over exchange rings. Isr. J. Math. **105**, 105–137 (1998)
5. Artin, E.: The influence of J. H. M. Wedderburn on the development of modern algebra. Bull. Am. Math. Soc. **56**, 65–72 (1950)
6. Atiyah, M.F., Macdonald, I.G.: Introduction to Commutative Algebra. Addison-Wesley, London (1969)
7. Beachy, J.A.: Introductory lectures on rings and modules. London Mathematical Society Student Texts, vol. 47. Cambridge University Press, Cambridge (1999)
8. Blyth, T.S.: Module Theory: An Approach to Linear Algebra. Clarendon Press, Oxford (1990)
9. Blyth, T.S., Robinson, E.F.: Basic Linear Algebra. Springer Undergraduate Mathematics Series. Springer-Verlag, London (1998)
10. Bourbaki, N.: Elements of Mathematics. Commutative Algebra. Springer-Verlag, Berlin (1989)
11. Canfell, M.J.: Uniqueness of generators of principal ideals in rings of continuous functions. Proc. Am. Math. Soc. **26**, 571–573 (1970)
12. Canfell, M.J.: An algebraic characterization of dimension. Proc. Am. Math. Soc. **32**, 619–620 (1972)
13. Clark, J., Lomp, C., Vanaja, N., Wisbauer, R.: Lifting Modules; Supplements and Projectivity in Module Theory. Frontiers in Mathematics. Birkhäuser Verlag, Basel (2006)
14. Conway, J.B.: A Course in Functional Analysis. Springer-Verlag, New York (1985)
15. Eilenberg, S., MacLane, S.: General theory of natural equivalences. Trans. Am. Math. Soc. **58**, 231–294 (1945)
16. Facchini, A.: Module Theory: Endomorphism Rings and Direct Sum Decompositions in Some Classes of Modules. Progress in Mathematics, vol. 167. Birkhäuser Verlag, Basel (1998)
17. Freyd, P.J.: Abelian Categories. An Introduction to the Theory of Functors. Harper & Row, New York (1964)
18. Gardam, G.: A counterexample to the unit conjecture for group rings. Ann. Math. **194**, 967–979 (2021)
19. Gillman, L., Jerison, M.: Rings of continuous functions. Graduate Texts in Mathematics, vol. 43. Springer-Verlag, New York (1976)
20. Golan, J.S.: The Linear Algebra a Beginning Graduate Student Ought to Know. Springer-Verlag, Dordrecht (2007)

© The Author(s), under exclusive license to Springer Nature Switzerland AG 2022
M. Mathieu, *Classically Semisimple Rings*,
https://doi.org/10.1007/978-3-031-14209-3

21. Hilton, P.J., Stammbach, U.: A Course in Homological Algebra. Graduate Texts in Mathematics, vol. 4. Springer-Verlag, New York (1971)
22. Jacobson, N.: Lectures in Abstract Algebra, vol. II. D. Van Nostrand, Toronto (1953)
23. Knapp, A.W.: Basic Algebra. Birkhäuser, Boston (2006)
24. Lam, T.Y.: A first course in noncommutative rings. Graduate Texts in Mathematics, vol. 131. Springer-Verlag, New York (1999)
25. MacLane, S.: Categories for the Working Mathematician. Graduate Texts in Mathematics, 2nd. edn., vol. 5. Springer-Verlag, New York (1978)
26. Mathieu, M.: Funktionalanalysis. Ein Arbeitsbuch. Spektrum Akademischer Verlag, Heidelberg-Berlin (1998)
27. McGovern, W.Wm.: Neat rings. J. Pure Appl. Algebra **205**, 243–265 (2006)
28. Nicholson, W.K.: Lifting idempotents and exchange rings. Trans. Am. Math. Soc. **229**, 269–278 (1977)
29. Osborne, M.S.: Basic Homological Algebra. Graduate Texts in Mathematics,, vol. 196. Springer-Verlag, New York (2000)
30. Pareigis, B.: Categories and Functors. Academic Press, New York/London (1970)
31. Passman, D.S.: The Algebraic Structure of Group Rings. John Wiley & Sons, New York (1977)
32. Pedersen, G.K.: Analysis Now. Graduate Texts in Mathematics, vol. 118. Springer-Verlag, New York (1995)
33. Taylor, M.E.: Linear Algebra. Pure and Applied Undergraduate Texts, vol. 45. American Mathematical Society, Rhode Island (2020)
34. Wallace, D.A.R.: Groups, Rings and Fields. Springer Undergraduate Mathematics Series. Springer-Verlag, London (1998)

Index of Symbols

M. Mathieu, *Classically Semisimple Rings*,
https://doi.org/10.1007/978-3-031-14209-3

Index

Printed in the United States
by Baker & Taylor Publisher Services